Mr. Lean and the Engine Reporters

Mr. Lean and the Engine Reporters

by

Bridget Howard

Trevithick Society

First published by The Trevithick Society Dec 2002

ISBN 0 904040 56 9

Frontispiece: Robinsons Engine, South Crofty Mine. Valve gear with Hardings Improved Counter, made by Harding Rhodes & Co, Leeds. (See page.84) Photo: South Crofty Collection
Cover picture: Valve gear of Taylors Engine, East Pool, with counter. Photo Tony Clarke

Printed and bound in Cornwall by R. Booth Ltd, Penryn, Cornwall, Great Britain TR10 9HH

CONTENTS

	Page
Introduction	7
Smeaton and Watt	9
Joel Lean	13
Whose Reporter ?	18
The Years of Conflict	24
John Lean's Reporter	31
The Influence of John Taylor	37
"Lean's Engine Reporter"	44
Change and Decay	48
Tonkin's Account	56
Browne's Cornish Engine Reporter	59
Eddy's Reporter	69
Taylor's Mexican Reporters	73
The Day's Work	76
Appendices	
1. The Measurement of Duty	81
2. Copies of the Reporters	86
3. References	89
4. Acknowledgements	94
Index	95

ILLUSTRATIONS

Page

1. Frontispiece: Counter on Robinsons Engine

2. Map of Cornwall   8

3. Smeaton's account of 15 large atmospheric engines operating
    in Cornwall in 1779. (Reproduced from Farey's *Treatise* by
    permission of David & Charles Ltd.)   10

4. Joel Lean's Reporter for December 1811.   14

5. Extract from the Leans' family tree.   16

6. Table showing the monthly average performances of certain
    engines, excluding Woolf's, August 1811 - May 1815.
    (Reproduced from the *Philosophical Magazine* 1815.)   26

7. Part of John Lean's Reporter, November 1827, as sent to
    the printer.   35

8. John Taylor, painted c.1826 by Sir Thomas Lawrence.
    Destroyed by fire 1870.   38

9. Title page of the *Historical Statement*.   41

10. Portrait of Thomas Lean II in Marazion Council Chamber.
    (Reproduced by permission.)   49

11. James Champion Keast. (Reproduced by permission of
    Mrs M. Bunney.)   54

12. William West of Tredenham, civil engineer. (Reproduced
    by permission of the Institute of Cornish Studies.)   60

13. Summaries by Lean and Browne of their respective
    Reporters, published in the *West Briton*, 17 September 1852. 66

14. Eddy's Reporter, September 1829.   70

15. Extract from Lean's Reporter, June 1841, showing the
    statistics for Taylor's engine at United Mines.   82

INTRODUCTION

Cornish steam engines were known throughout the world, and used wherever a pumping engine of great power and reliability was needed. What made them unique was not merely their design, but also the meticulous records that were kept of their performance, so that intending purchasers could weigh the advantages of one engineer's model against that of his competitors. These statistics were assembled throughout the 19th century, month by month; they measured the work done by the engines in the mines of Cornwall, and scientifically compared their efficiency. These records were known as Engine Reporters.

The best known were compiled by the Lean family from 1811 until 1904. But there were others. A breakaway version was issued by John Lean, brother of the then editor, and lasted from 1827 to 1831. A rival text appeared under the name of William Tonkin from 1834 for about eight years. William Browne's was published between 1847 and 1858. Two further series were written for John Taylor, giving him information about the engines at his mines in North Wales and in Mexico.

The reporters used a measurement known as 'Duty' to compare the efficiency of the engines. Duty was the result of dividing the amount of work done by the quantity of fuel used. It was expressed as millions of tons lifted a foot high by burning a bushel (later a hundredweight) of coal. The higher the duty, the better the performance. The highest duty recorded was 125 million at Fowey Consols in 1835.

But these figures were calculated by men with human failings. Today, we assume that the truth is exactly as set out in the dry statistical columns. Usually it was, but we forget that mining and engine building were fiercely competitive industries. Those who compiled the reports were involved in the conflict, particularly in the early days. They had their friends and their feuds. The elder Thomas Lean was a rogue; his son was a pillar of Victorian rectitude. William Tonkin was a very dubious character, and William Browne is still a figure of controversy. We cannot ignore the personal histories of these men without ignoring an important aspect of the history of Cornish technology. This book tells their individual stories.

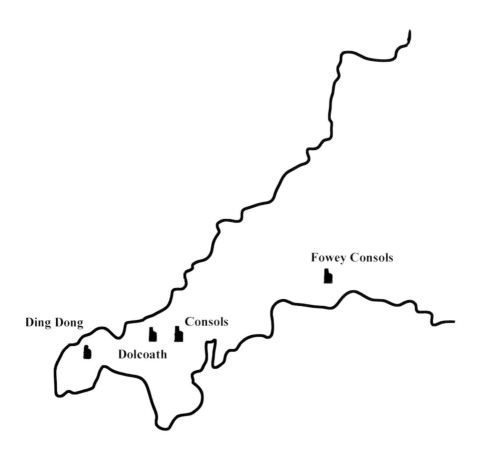

Ding Dong

Dolcoath

Consols

Fowey Consols

SMEATON AND WATT

The purpose of Lean's Engine Reporters was to provide a regularly published series of statistics from which the efficiency of each engine could be calculated, and thus compared with the others. Although these Reporters were different from anything that had been seen before, they were not the first analyses of the work done by Cornish engines.

The earliest published tabulation of the expected (not, as Lean's, the actual) performance of steam engines appeared in 1719. It was prepared by Henry Beighton and was printed in *The Lady's Diary*[1]. This followed the introduction into Cornwall by about 1716 of the engines designed by Thomas Newcomen (1633-1729). But these used a great deal of fuel and all coal for the steam engines had to be brought in by sea. This was expensive. Until 1831, the Cornish (unlike the industries up country) had to pay a tax on sea-borne coal, in addition to transport and shipping costs, and on top of the price of the coal itself[2]. It was therefore vital to know, and to minimise, the amount of fuel that each engine consumed. For decades, it was the cost of coal that inspired the improvements in the performance of the Cornish beam engines.

The first person to devise a method of calculating efficiency in terms of coal consumption was John Smeaton (1734-92). In 1772, he compiled a 'Table for the Proportions of the Parts of Newcomen's Engines'. This constructed the optimum characteristics of 32 idealised engines, and computed the quantity of water that would be raised one foot per minute, and also the effect per minute of burning a bushel of coal per hour[3].

Seven years later, in 1779, his calculations turned from the theoretical to the actual, monitoring the performance of 15 large Newcomen engines working at Dolcoath, Poldice, and at Wheals Virgin, Maid and Chance. He assumed, probably incorrectly, that the engines were working continuously, and took the fuel consumption figures from the accounts showing coal tax rebates. Inevitably, his results were inaccurate, but gave an average duty of 7.10 million for 12 of the 15 engines, with the worst showing 5.30m and the best 7.81m: low in comparison with the cylinder sizes. No coal consumption amounts were available for the three engines at Dolcoath, and their duty could not be assessed[4].

About 1775, Cornish mining interests made their first tentative approach to James Watt (1736-1819), perhaps not realising that the wording of his patent could prohibit any steam engines in the county other than his

Performance of Fifteen Atmospheric Engines working at Mines in Cornwall, in 1779.

Names of Mines.	Horse power exerted.	Diam. of cylinder. inches.	Area in square inches.	lbs. square inch.	Motion per min. feet.	Millions lbs. per bushel.	Bushels per month.	Depth drawn. fathoms.	Cubic feet drawn per minute.
Wheal Virgin.	18·80	60	2827	5·70	38·5	5·82	4670	35	47·35
	18·25	60	2827	5·54	38·5	5·30	4980	34	47·35
	22·08	60	2827	7·00	38·5	7·79	4284	43	47·35
	20·45	64	3217	7·63	27·5	7·47	3960	71	25·35
	23·65	70	3848	6·60	38·5	7·63	5616	43	60·80
Poldice.	14·78	60*	2827	5·23	33·0	7·17	2979	25 }	52·20 }
	19·51	66*	3421	5·70	33·0	7·26	3945	33 } 58	52·20 }
	15·46	66	3421	5·42	27·5	7·81	2864	63	21·60
Wheal Maid.	28·28	60	2827	6·66	49·5	6·76	6048	36	69·30
	25·13	65	3318	5·05	49·5	7·42	4896	32	69·30
Dolcoath.	17·88	60	2827	6·32	33·0	----	----	42	49·20
	16·65	63	3117	5·88	30·0	----	----	63	29·80
	20·93	70	3848	5·98	30·0	----	----	71	37·56
Wheal Charce.	16·02	69¼	3794	6·97	20·0	7·27	3186	51	28·00
	16·66	70	3848	7·14	20·0	7·56	3186	56	28·00
Averages.	20·04	64¼	3253	6·18	33·8	7·10	4210	46½	44·30

1. Smeaton's account. Manuscript notes are by John Farey

own. For each engine, he received a payment equal to one-third of the value of coal saved, based on the performance of two Newcomen engines at Poldice, assessed in August and September 1778. These used 14,080 bushels of coal in 61 days, and were judged to have given a duty of 7.037m[5]. (In the following year, Smeaton's tests gave 7.80 and 7.20m.) Various adventurers subsequently grumbled that Watt's standard should have been set against one of Smeaton's engines instead of against old atmospheric machines, but by then it was all too late. Watt's engines were fitted with specially designed counters that recorded the number of strokes made, thus enabling the duty (and the premium) to be estimated exactly.

Following Watt's *Calculation and Performance of Cornish Engines 1778*, his partner, Matthew Boulton (1728-1809), kept his own monthly statistics, taking the record forward from 1779 to 1782. The pair carefully preserved their notes on their own and on the existing Newcomen engines in case of any future disputes, and Watt's Blotting and Calculating Books contain the results of tests, measurements of performance, and the fees accounts.

At the height of their dispute with Jonathan Hornblower, two papers were published under the name of Thomas Wilson (1748-1820), the partnership's commercial agent in Cornwall. The first was *A Comparative Statement of the Effect of Messrs Boulton and Watt's Steam Engines with Newcomen and Mr Hornblower's* (issued in 1792; 25 pages). The second, a year later, was *An Address to the Mining Interests of Cornwall on the subject of Messrs Boulton and Watt's and Hornblower's Engine* (22 pages). Davies Gilbert (1767-1832), the Cornish technocrat and adviser to Hornblower, found it full of misrepresentations and based on the inaccurate tests that had been made on the latest Watt engine[6]. It was propaganda rather than scientific research, and was intended to bolster Watt's reputation.

The performance of his engines declined over the next five years, and, following disputes about the royalties, an arbitration committee was appointed in 1798 with Davies Gilbert as adjudicator. He calculated the duty of each of the 23 engines in question, and found that the average result was 17.7m, with the worst engine giving 6m and the best 27.5m. The locations of the engines are not given in the paper that Gilbert presented to the Royal Society in 1830[7], but the best was at Herland, where the agent, Captain John Davey, said that it usually did 20m[8].

With the expiry of the patent in 1800, Boulton and Watt left Cornwall. It is a moot point whether they had benefited the county, but for

11

them it had been a gold mine, netting nearly £180,000 in fees. With their departure, as Matthew Loam recalled, a dark age began, when the Cornish engine 'had dwindled to a filthy jumble of a thing ... with steam flying in all directions'[9]. Thomas Lean spoke of the county being bereft of men who were capable of maintaining the engines and

> a great and general deterioration of the machines, which continued for many years; until the duty performed had sunk so low as would appear almost incredible to men acquainted with those of the present [i.e. 1838] day[10].

It was probably the publication in 1811 of what became Lean's Engine Reporter that saved the situation and gave the industry the kick-start that it needed.

JOEL LEAN

Joel Lean (1749 - 1812) is generally believed to have been the man who started the Reporters. An engineer, a mathematician, an innovator, a Methodist of transparent integrity, he was the ideal candidate for the job, particularly since at the age of 62 he had no further career ambitions. A search throughout Cornwall could have found no better editor, no one more likely to command the respect of all the mining industry. How providential that his was the name on the new Reporters, and that his spotless reputation was behind them.

He was born at Gwennap, south-east of Redruth, the son of William and Araminta Lean, an intellectual couple who were early converts to the teaching of John Wesley. They brought up their children to have the highest standards of personal integrity and encouraged them to make the fullest use of their God-given abilities. Young Joel became a miner, but his hobby was arithmetic and by about 1775, when he was in his mid-20s, his mathematical puzzles were regularly published in *The Gentleman's Diary*. At the same time, Miss Blanche Harris, daughter of Francis Harris of Mylor, was contributing poems and conundrums to *The Lady's Diary*. She had already received recognition, both as a serious poetess and as compiler of her Enigmas, which, with their answers, were in verse. The two began writing to each other, were persuaded by their friends to meet, and were married on 2 January 1777, 'shaking hands for life' as Blanche told her readers[11].

They both continued publishing puzzles until 1785, by which time Blanche had three sons and Joel was the mathematical correspondent for both Diaries. His wife continued to write occasional verse, including an 80-line elegy 'Reflections on sublunary enjoyment' composed while they were in Spain in 1806.

The couple lived first at Gwennap, moved to Ludgven in the early 1780s when Joel's work took him to that area, and finally settled in the village of Crowan, between Camborne and Helston, when he was appointed manager of the nearby Crowan and Oatfield mines in about 1801: a job that lasted until 1806. During his time there, during the decade after the expiry of Watt's patent, Cornish engineering seemed to be in terminal decline, but new growth was stirring among the ashes, and Lean may have been one of its progenitors. Someone, perhaps he, perhaps John Davey (1757-1819), perhaps the young Richard Trevithick (1771-1832), put

MINE	ENGINE	Load per square inch in Cylinder	Length of stroke in Cylinder	No. of Lifts	Depth in fathoms	Diameter of Pump in inches	Time, 1811.	Consumption of Coal in Bushels	Number of Strokes	Length of stroke in Pump	Load in pounds	Pounds lifted one foot high by a Bushel of Coal	2½ lengths per Minute	REMARKS.
		lb.	ft. in.		fm. ft.	in.				ft. in.				
Wheal Alfred.	Fast 60 inches Single.	8¼	7 3	1 2	16 4 / 46 2 / 22 3 / 8 0	9 / 13 / 10 / 9	December 31st	3168	325880	7 6	21704	19,030,042	7. 3	Drawing perpendicular 66 fathoms, underlays on the load 56 fathoms, with Bob over the Cylinder
Ditto.	Middle 66 Single.	7½	8 4	1 1 2 1	16 2 / 26 5 / 59 5 / 4 3	10 / 11 / 17 / 10	December 31st	3528	678560	6 6	31120	17,521,529	6. 8	Drawing perpendicular with Bob over the Cylinder
Ditto.	West 64 Single.	8½	8 4	3	17 3 / 72 0	10 / 14½	December 31st	3510	330580	6 6	31542	21,146,100	7. 4.	Drawing perpendicular with Bob over the Cylinder
Dolcoath.	Great Engine 63 Double.	14½	7 6	8 2	138 1 / 19 0	12 / 13	1812 January 2nd	7110	311950	7 6	47208	14,819,461	7.	Drawing perpendicular 158 fathoms, on the underlay 19 fathoms, with Cylinder over the Shaft, and Bob over the Cylinder; and three Balance Bobs in the shaft.
Ditto.	Shamrock 45 double, working single.	7½	8 2	9 1	39 4 / 12 0	13 / 5	January 2nd	1294	638580	8 0	11359	14,362,920	5. 8	Drawing perpendicular with Bob under the Cylinder.
Ditto.	Stray Park 63 Single.	7½	7 1	4 2 1	80 2 / 43 0 / 8 0	11½ / 11 / 7	January 2nd	3174	299910	5 4	32240	11,845,562	6. 7	Drawing perpendicular 123 fathoms, underlay 8 fathoms with Bob over the Cylinder.
Cook's Kitchen.	96, Double, working single.	9	7 0	5 1 1 1	82 0 / 9 0 / 3 3 / 3 0	10 / 9 / 6 / 5	January 2nd	1954	338220	3 0	18070	13,920,346	7. 8	Drawing all the load with flat rods, 40 fathoms perpendicular, 57 fathoms on the underlay, and 20 fathoms flat rods under ground
Wheal Fanny.	58 Double.	14	7 6	1 3 1 1	21 0 / 63 0 / 3 0 / 3 0	14½ / 14½ / 11 / 6	January 2nd	6192	406170	7 6	37400	18,399,658	9. 1	Drawing perpendicular with Cylinder over the shaft, and Bob under the Cylinder.

2. Joel Lean's Reporter; 2nd January 1812 (headed 1811 because most statistics are for 1811).

forward the idea that the old bucket pumps should be replaced by the plunger.

Lean did convince his adventurers that they should change to plunger poles and replace two small engines with a single 70-inch, at the time the biggest working in the county. As a consequence, coal consumption was reduced by half, the engines no longer broke down, output increased and financial liability became a profit. But this happy situation did not last for long. In the autumn of 1806, a quarrel broke out, which was to last for five years, between the proprietors of Crenver and Oatfield and those of the neighbouring Wheal Abraham. The mines closed and Lean was out of work.

Having been so successful in improving his own engines, he tried to persuade the managers of other mines that it would be in their interests to modernise their own pitwork and keep the engines in good repair. Most of his advice fell on stony ground. Copper prices were fluctuating, with many mines in financial trouble; exports were banned because of the war against Napoleon, but substantial amounts of foreign copper were being imported. Cornish engineering had reached its nadir: Hornblower had been bankrupted by his suits against Watt; Trevithick had gone to London; Bull had died; no one seemed to be interested either in developing new engines or in improving the old ones. Yet, in June or July 1811, there appeared the first of what was to become *Lean's Engine Reporter*, a monthly publication that lasted for almost a century.

The first issue was a trial and only gave details of the performance of three engines at Wheal Alfred in the preceding month. It was in tabular format, on a single folio sheet, with the name of Joel Lean printed in large capitals at the end. The manager at Alfred was Captain John Davey and the engines were reckoned to be the best in Cornwall, giving an average duty of about 20 million[12]. This Reporter seems not to have been made generally available, and the second issue, published in September 1811 with data for August, is usually taken to be the first in the series. It was headed 'Work done by the following engines for the last month' and included eight engines. By March 1812, sixteen were being covered, and the average duty had dropped to 15,770,000[13]. When Joel died, in September 1812, eighteen were being listed, with the highest monthly performance so far of 19,355,000[14]. A separate Whim table (headed 'Work done by the following engines drawing the stuff out of the mines') began in May 1812 and continued throughout the life of the Reporters.

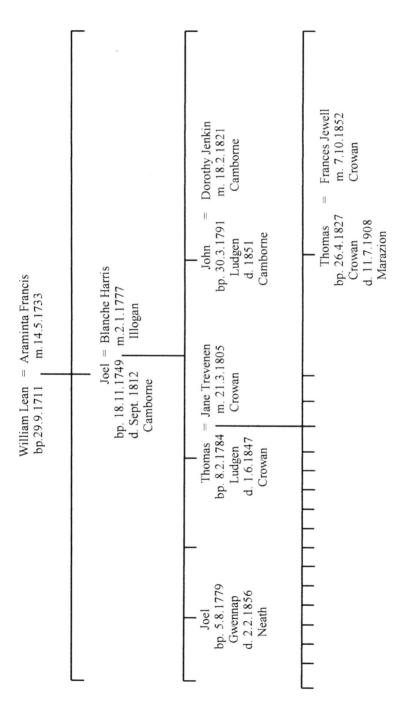

William Lean = Araminta Francis
bp.29.9.1711   m.14.5.1733

Joel = Blanche Harris
              m.2.1.1777
bp. 18.11.1749   Illogan
d. Sept. 1812
Camborne

Joel
bp. 5.8.1779
Gwennap
d. 2.2.1856
Neath

Thomas = Jane Trevenen
bp. 8.2.1784   m. 21.3.1805
Ludgen   Crowan
d. 1.6.1847
Crowan

John = Dorothy Jenkin
         m. 18.2.1821
bp. 30.3.1791   Camborne
Ludgen
d. 1851
Camborne

Thomas = Frances Jewell
             m. 7.10.1852
bp. 26.4.1827   Crowan
Crowan
d. 11.7.1908
Marazion

3. Extract from the Lean family tree

16

Perhaps their most innovative feature was that they were issued at regular intervals over a long period, rather than being a snapshot of performance either on a given day or as an annual average. Lean revealed how an engine was working, month by month, giving a record of what was actually happening. Furthermore, it was not just the chosen few that were monitored; it was open to the management of any mine that cared to pay the subscription to have their engine listed.

The Reporters were something entirely new. Setting them up had required both imaginative vision and hard work. Joel was now past his prime, and it was surprising that he should have bothered with this new project, however enthusiastic he had been about improving engine performance. The job would be physically demanding and was unlikely to make him a great deal of money. At his time of life he should have been ready to retire, instead of embarking on a new career. He seems an unlikely man to have initiated the publications.

WHOSE REPORTER ?

Thomas Lean the elder

When Joel died in September 1812, his fourth son, Thomas (1784-1847), and his fifth son, John (1791-1851), took over the work of compiling the monthly reports. The name of Lean appears on all issues and we therefore assume that the publication originated, was owned and managed by successive generations of the family. But there are indications to the contrary: indications that the Leans did not start the Reporter, nor, for nearly 30 years, did they control it.

It was Thomas who let the cat out of the bag in 1839 when he and his older brother, Joel junior (1779-1856), wrote their *Historical Statement of the Improvements made in the Duty performed by Steam Engines in Cornwall from the Commencement of the Monthly Reports*[15]. Joel's name did not appear in the book. The title page gave the authors as 'Thomas Lean and brother', and it is clear that Thomas was taking the initiative. He described how his father had wanted his fellow mine managers to publish their duty figures, but they had refused. Years went by, and

> the first who seems to have been sufficiently alive to the subject was the late Captain John Davey of Gwinnear, who was the principal manager of Wheal Alfred: and accordingly the first report was published in 1811[16].

The name of Captain William Davey was also mentioned. This was the first time that this story had been told and it seems odd that Thomas Lean should have waited nearly 30 years (1811-1839) to reveal the Daveys' involvement.

Another curious comment in the same passage is Thomas' praise of the Reporters, whose publication, he said, had been 'attended with more benefit to the county than any other single event except the invention of the steam engine itself.' Bearing in mind that the Lean's name had been the only one on the Reporters since the beginning, this is a remarkably immodest claim. It only makes sense if it was intended as a tribute to someone else. Who? Was it correct that the Daveys had initiated the Reporters?

What had really happened in the summer of 1811? This was when the engineer Arthur Woolf (1766-1837) returned to Cornwall. In the August, the first Reporter was issued. The two events coincided. Twenty-six years later, in 1837, there was another coincidence. Woolf died and Thomas Lean soon began writing the *Historical Statement,* taking the

18

opportunity to break the news that the Daveys were responsible for starting the Reporter. It is stretching credulity too far to believe that all these were fortuitous. Two such links with Woolf must be significant.

Why should the Daveys have set up the Reporters? However strongly they believed that engineering standards should be raised, what was in it for them? Why should they wish to encourage competition with their mines? What thanks would they have received from their masters? Woolf, on the other hand, had every incentive. He wanted to sell his engines in Cornwall. He needed to be able to prove that they were the best. He wanted to make a fortune from the premiums as Watt had done. For him, the Reporters were vital; for the Daveys, they would have been an altruistic gesture.

Arthur Woolf

Woolf was born in 1766, the son of a millwright at Dolcoath, served his apprenticeship to a local carpenter and then went to London to seek his fortune. He found employment with Joseph Bramah (1749-1814), the future famous locksmith, became a skilled metal worker and turned to engineering. He moved to the Griffin brewery in Clerkenwell and, after successfully installing a new engine for them, began to experiment in the design of steam boilers, taking out the first of his four patents in 1803. He fell out with the Griffin, set up in partnership with Humphrey Edwards (a millwright) and together they opened an engine works in Lambeth. He made the acquaintance of Alexander Tilloch (1759-1825), proprietor of the *Philosophical Magazine*, who funded his experiments, paid the expenses of obtaining his patents, publicised him in his magazine[17], and introduced him to many of the great men of the day. The heady excitement of all this decided Woolf to return to Cornwall to make his fortune there, as Watt had done. He laid his plans, and, just before leaving London, terminated his agreement with Edwards in May 1811. His tactics had been carefully worked out: he needed Cornwall and he was going to make sure that Cornwall needed him.

The advertisement that he placed in the *Royal Cornwall Gazette* on 16 February 1811 was the first visible sign that he intended returning to his birthplace. This announcement told the mining community that his patent steam engine had been brought to such perfection that he guaranteed it to require only half the fuel used by a Boulton and Watt engine; his customers would pay a premium of one-third of the cost of the coal that they saved. On the succeeding 17 May, there was a similar advertisement in the *West*

Briton. Later that month, by which time he had moved to Cornwall, he had a meeting at Wheal Unity with Captain William Davey at which Alexander Tilloch was present. In the July, John Davey began collecting the duty information on the Wheal Alfred engines which was published in August as the preliminary issue of the Reporter.

The connection between Woolf's return and the starting of the Reporter is made quite clear in an article in the *Philosophical Magazine* in 1815, possibly written by Thomas Lean and certainly approved by Woolf. Being published in a London journal that was not widely read in Cornwall, the statement does not seem to have attracted any local reaction. After praising Woolf's engines, the writer continued

> some of the ends intended by the monthly reports of work actually done by the engines employed in the mines, particularly in the pumping, was to know the comparative merits of Woolf's engines with two cylinders when compared with steam engines in general use[18].

Woolf's other particular friend, the civil engineer John Farey (1791-1851), agreed that the Reporters were started so as to 'have regular accounts kept of the actual performance of his engines'[19]. It is clear that the purpose of the Reporters, and the reason for setting them up, was to demonstrate how much better Woolf's engines were than any of the others. Any encouragement that they may have given to the raising of general engineering standards was unintentional. The Reporters were skilfully constructed advertisements issued under the guise of impartial statistics.

It may be thought that there was insufficient time between Woolf's return and the beginning of the Reporters; surely, he could not have been involved? He was back in Cornwall in May 1811, statistics were assembled at Wheal Alfred in July, the first Reporter was published in August, and the series began in earnest in September. This gestation period seems much too short. In reality, however, the start-up programme was carefully planned before being carried out. It was not a rush job.

If we look behind the scenes, there was obviously a great deal of preliminary thinking and planning. The monthly Reporters were innovative; nothing like them had been seen before. They were not simple documents. There had to be consensus about their content and their format, about the information which was to be revealed, the methods of collecting the data and then displaying it. Fiercely independent mine managers and adventurers had to be persuaded not only to participate, but also to pay for that pleasure. Then there were the counters that had to be fitted to every engine before it could be listed. Such instruments had not been seen for a decade. New ones had to be designed, made, and a supply of spares kept ready for future use. All this could not be done in a few weeks. There was engineering and milling to be done in London; there was persuading and conciliation to be carried out in Cornwall. Work for Woolf and probably for the Leans and the Daveys.

Everything points to the Reporters having been planned well before Woolf's arrival in May 1811. The first newspaper advertisement in the February of that year was the signal that the scheme was ready. Woolf came to Cornwall with the purpose of using the Reporters to promote himself and his engines, bringing with him a supply of the essential counters which he had made in his Lambeth workshop and which had been fitted with locks made by his friend and ex-employer Joseph Bramah[20]. The Reporters were not a consequence of Woolf's return: they were essential to it.

Once the strategy had been agreed, there was need of someone to handle the day-to-day routines. This had also been planned well in advance. As long before as August 1810 Woolf's friend Andrew Vivian was offered a salary of £150 a year to prepare the Reporters, but had refused on the grounds of being too busy. (By way of making amends, Vivian who was manager of Wheal Abraham appointed Woolf as engineer there immediately he returned to Cornwall.) Vivian suggested Richard Trevithick as reporter, but he too refused[21]. The third choice was Joel Lean.

This hawking around of the job makes it quite clear that Lean was not the instigator of the Reporters, merely the agent of other interests. According to the *Historical Statement*, he was approached by the Daveys.

> While Captain Davey had the matter on his mind ... he happened to be conversing on the subject with the late Captain William Davey of Wheal Unity; when one of them remarked 'That such a thing is most desirable cannot be denied, but where shall we find a man possessing the necessary qualifications: leisure to attend to it; a thorough acquaintance with the business; and, above all, known integrity in which the public will be willing to confide?' The other, looking up, saw Captain Lean coming towards them, and replied, 'Here is the very man we want, possessing all the qualifications you require.' He was accordingly requested to take upon him the

office of Registrar and Reporter[22].

The only problem was that he was now 62. The solution was for his sons Thomas and John to do the legwork, and for Joel to supervise them. It was he therefore that signed the Reporters, underwriting them with his good reputation. This continued for 14 monthly issues, until he died in September 1812.

Thomas and John then began signing the Reporters themselves, but, freed from Joel's restricting influence, Woolf started to manipulate things, perhaps providing the technical expertise lacked by the Leans, and, most certainly, bullying them into doing what he wanted. John Lean eventually rebelled, claiming that incorrect duty figures were being given in the Reporters. Thomas seems to have had fewer scruples about being made use of to publicise Woolf's engines, and even gave advance information about their performance before it appeared in the Reporters[23].

Their publication continued amid charge and counter-charge of dishonesty. From 1818 to 1830, Woolf was employed by the mining entrepreneur John Taylor (1779-1863) as engineer at the Consolidated Mines. During the 1830s, when Woolf no longer worked for him, Taylor wrote paper after paper defending the integrity of the Reporters. Woolf died in October 1837 and almost immediately Taylor paid the Leans to write the *Historical Statement*, ostensibly a chronicle of the Reporters, in reality another defence of Cornish engineering which had been brought into so much disrepute.

The book is a skilful presentation, mixing spin and fact, encouraging conclusions to be drawn from incomplete evidence. Without departing from the truth, it does not always reveal the whole truth. It purports to be a uniquely authentic record of Cornish engineering. Today its accuracy is (almost) unchallenged, and the dust-jacket of the 1969 reprint speaks of "the faithful charting" of technological progress "in Lean's columns". There is of course no mention of Woolf's involvement with either the Reporters or the Leans. The *Statement* is powerful propaganda.

In it, Lean produced the tale about the Daveys having initiated the Reporters. By then, the brothers, like Woolf, were dead. Taylor, in commissioning the book, would have kept a close eye on the text, and presumably approved it. Telling the story in such a prestigious publication prevented any other awkward questions about the origin of the monthly reports. There was no pretence that the Leans had been responsible, and the Daveys from six feet underground could not deny what was said. They were given the posthumous credit by Taylor and the Leans, who all knew what had really happened. In death, as in life, the Daveys lent bogus respectability to Woolf's business plan.

THE YEARS OF CONFLICT, 1812-1831

Thomas and John Lean took over the Reporters in the second half of 1812: no copies exist to prove in which month. By January 1813, they had made one significant change. The name of the superintending engineer was now given at each mine.

For the next two years, the most successful engine was the 63-inch at Stray Park. This headed the duty lists during 1813, doing 35 million in February 1814 and 33.8m in the March, so far exceeding all the others as to cause the Leans to add a footnote to the April 1814 Reporter to allay rumours that 'some trick had been used to make her appear better than she really is'. (The belief was that she was being run on a very much shorter stroke than was being stated.) Her dominance was ended by Woolf's new engine at Wheal Abraham which was listed as doing 34m from September to November 1814. Not content with this, Woolf ran a private 24-hour trial and claimed 70m duty[24].

In May 1812, a separate table for winding engines had appeared in the Reporters; in May 1813, the first of Woolf's patented engines for winding was started up at Wheal Fortune and was noted as "Mr. Woolf's engine": none of the others gave their builder's name. A comparable table for stamps began in 1817, but relatively few of them appeared in any monthly part and the series ended in December 1851.

Lean himself was no engineer, but he had been one of the official adjudicators at the public trial of Woolf's whim engine at Wheal Abraham in October 1813. His lack of technical knowledge came as a surprise to some visitors to the mines, among them Thomas Coulson (from the Royal Dockyard at Devonport) who was making enquiries some years later on behalf of Simon Goodrich of the Navy Board. He sent a copy of the December 1829 Reporter to his friend at Portsmouth Dockyard, and remarked

> The author of this report ... could not furnish me with drawings or particulars as I expected nor do I think that any things that he says that I collect would be of sufficient accuracy for your guidance[25].

Nonetheless, in spite of his lack of technical expertise, Lean was asked to give evidence in May 1817 before a Select Committee of the House of Commons that was investigating the causes of boiler explosions, and the comparative safety of cast- and wrought-iron boilers using high pressure steam. The enquiry was set up following an accident to a steam-driven passenger ship. Engineers from around the country were called, including from Cornwall, Andrew Vivian, concerned that 'the legislature would not interfere, as to prevent the use of high-pressure steam engines', and Arthur Woolf, who explained the perfections of his patent engines. He stated that all the explosions that had occurred in Cornwall had been in wrought-iron boilers, using low-pressure steam. In his enthusiasm to make his case, he said that he would not hesitate to seat himself on an exploding cast-iron Cornish boiler, because the damage would only take place in its lower parts[26].

To mark the starting of Woof's new 53-inch at Wheal Vor, Tilloch published a paper in the 1815 *Philosophical Magazine* entitled "Some Observations on Steam Engines; with a Table of Work done by certain Engines in Cornwall". This showed, for each month from August 1811 to May 1815, the number of engines in Lean's Reporter and also tabulated the bushels of coal consumed each month by all the engines; pounds of water lifted one foot high by all the coals consumed; and the average of pounds lifted one foot high with each bushel of coals[27].

The figures, however, were deliberately misleading. The clue lay in the title to the article: 'Work done by certain engines.' The 34 engines that were listed as performing an average duty of 20.4m in May 1815 did not include Woolf's new 53-inch at Vor. That had done 49.9m in the month, and to have included it would have taken the general average up to 35m. It was much more effective publicity to demonstrate that the other engines were all doing the same kind of duty that they had done for the past two years, whereas Woolf's new machine on its first listing in the Reporter had more than doubled the average. There was also advance information.

> ... by letter we are informed (for the printed Report has not yet reached us) that the duty performed by Woolf's engine in the month of June was 50,333,000.

To avoid any misunderstanding, the table (updated to November 1815) was reprinted, with further explanations, in the December issue of the *Magazine*[28]. This lauded the duty of 50m done by the engine at Wheal Vor in the November, and mentioned the 52m performed by Woolf's engine at Wheal Abraham during the same month.

The first of these two articles is credited by the authoritative

TABLE.

	Number of Engines reported	Bushels of Coals consumed by all the Engines	Pounds of Water lifted one foot high by all the Coal consumed.	Average of Pounds lifted one foot high with each Bushel of Coals.
1811. August - -	8	23,661	126,126,000	15,760,000
September -	9	25,237	125,161,000	13,900,000
October - -	9	24,487	121,910,000	13,510,000
November -	12	30,998	189,340,000	15,770,000
December -	12	39,545	204,907,000	17,075,000
1812. January -	14	50,089	237,661,100	16,972,000
February -	15	54,349	260,511,000	17,900,000
March - -	16	59,140	274,222,000	17,188,000
April - -	16	62,384	276,233,000	17,260,000
May - -	16	51,903	275,546,000	17,096,000
June - -	17	50,410	238,076,000	16,940,000
July - -	17	51,574	300,441,000	17,677,000
August - -	17	44,256	314,753,000	18,510,000
September -	18	46,526	348,396,000	19,355,000
October -	18	55,911	521,990,000	17,883,000
November -	21	57,176	381,460,000	18,160,000
December -	19	55,784	541,303,000	18,200,000
1813. January -	19	60,400	363,906,000	19,155,000
February -	22	58,044	438,737,000	19,940,000
March - -	23	73,862	440,612,000	19,157,000
April - -	23	61,739	431,032,000	18,760,000
May - -	24	58,890	463,346,000	19,300,000
June - -	24	53,110	470,157,000	19,590,000
July - -	23	56,709	443,462,000	19,281,000
August - -	21	50,110	416,898,000	19,852,000
September -	22	58,008	427,148,000	19,113,000
October -	26	74,796	488,671,000	18,795,000
November -	28	77,135	537,958,000	19,212,000
December -	29	86,273	584,721,000	20,162,000
1814. January -	28	91,753	550,751,000	19,670,000
February -	26	78,986	536,677,000	20,641,000
March - -	23	109,904	565,465,000	20,193,000
April - -	29	91,607	576,617,000	20,525,000
May - -	28	79,437	569,319,000	20,305,000
June - -	30	75,343	626,669,000	20,888,000
July - - -	27	85,224	573,208,000	21,222,000
August - -	26	70,443	545,019,000	20,961,000
September -	27	78,167	560,608,000	20,763,000
October -	32	75,030	630,704,000	19,709,000
November -	32	82,000	637,322,000	19,916,000
December -	29	84,669	573,744,000	19,784,270
1815. January -	32	110,824	637,320,990	19,916,250
February -	33	101,667	710,271,250	21,525,370
March - -	34	117,342	706,071,990	20,766,820
April - -	35	105,701	695,212,340	19,863,210
May - -	34	107,530	669,299,140	20,479,350

4. Table in "*Philosophical Magazine*" showing the monthly average performance of certain engines, excluding Woolf's.

Bibliotheca Cornubiensis to Thomas and John Lean, but William Henwood thought otherwise. He sarcastically commented that

> Woolf and Tilloch by their frequent publications on the subject, industriously propagated the opinion of this superiority[29].

The December issue of the *Magazine* included a letter from Joseph Price of Foxes' and the Neath Abbey Iron Company, stating his belief that Woolf's engines were 'by far the most eligible where a saving in the consumption of coal is an object' since they performed to the same duty as good Boulton and Watt engines, with less than half the quantity of fuel. This testimonial was followed by a certificate of performance, signed by the principal agents of Crenver, Oatfield and Abraham mines, by Thomas Lean, and by William Davey who annotated his signature by commenting: I occasionally attend at Wheal Abraham, and believe the above statement to be true. (His brother John was manager there, but did not sign, and Woolf was the superintending engineer.) This was followed by a similar document signed by the agents at Wheal Vor (engineer Arthur Woolf) but not by either Davey. The first certificate estimated the coal saving as against Watt's engines as 34:20, and in the previous four months as 47:20. At Wheal Vor, the saving was reckoned to be 116:34[30]. Readers of the *Magazine* may not have known of the business relationship between the Daveys and Woolf, nor realised that Joseph Price's ironworks were building Woolf's engines. Richard Trevithick, who had fallen out with Price, threatened to horsewhip him 'for the falsehoods that he with the others had reported'[31].

An interesting contemporary sidelight on the *Magazine* articles can be found on a copy of the June 1818 Reporter in the Goodrich collection in the Science Museum. An unknown writer, possibly at the Neath Abbey ironworks, who were the recipients of this particular copy, had taken the trouble to compare the duty figures in that issue of the Reporter with those of the engines cited in the *Magazine* for May-June 1818. He notes 'These accounts agree with the printed Report'. Judging from the pen and ink sums scattered on its printed pages, he had also deemed it necessary to make his own computation of the duty figures.

The *Philosophical Magazine* continued to give publicity to Woolf, for example,

> From Messrs Leans' Report for May 1818, it appears that during that month, the following was the work performed by the engines reported with each bushel of coals[32]

	Pounds of water lifted 1 foot high with each bushel	Load per square inch in cylinder in lbs
23 common engines averaged	23,608,329	various
Woolf's at Wheal Vor	29,032,182	17.2
ditto Wheal Abraham	31,520,346	16.8
ditto ditto	29,702,703	5.68
Dolcouth [sic] engine	38,233,193	11.2
Wheal Abraham ditto	33,714,842	10.9
United Mines engine	33,967,127	13.6
Treskirby ditto	40,615,253	10.8

Comparison with the Reporter for May 1818 shows that the engines at the United Mines and at Treskirby were commonly each known as Sim's Engine (which Tilloch chose to forget), and also that the engine at Cook's Kitchen and another at United Mines had done 30 and 32 million respectively — higher than the Woolf's engines that were highlighted — but were also conveniently forgotten.

The importance of the duty figures to Woolf is illustrated by events at the public trial of his Wheal Abraham engine in August 1818. This compound (together with that at Wheal Vor) had consistently headed the ratings, but had just been refitted so as to ensure that it was in prime condition to beat the competition that was expected from new engines at Dolcoath and at Wheal Chance, both due to start in September 1818. At the trial, the Wheal Abraham engine produced a duty of 65.21 million, the highest yet officially attained[33].

John Farey, who was himself a consulting engineer, attended the trial and took the opportunity of the refit to measure the engine's component parts. He found that, although the correct dimensions had been used to calculate the results of the trial, inaccurate figures were quoted in the Reporter. The total weight of all the columns was said to be 25,826 lbs, instead of 24,955 lbs, and the length of the stroke was not the alleged 7 ft but was $6^2/_3$ ft, thus enhancing the duty[34]. Writing later, at the end of the 1840s, Farey made the comment that the monthly reports were not then [i.e. in 1818] 'so well established as they were afterwards'[35].

There were rumours at the time that the extraordinarily high performance had not been obtained by honest means[36] and, in view of events nine years later, some doubt must attach to others of Woolf's duty figures, possibly throughout this period. His desperation for success is understandable because his patent expired that year and his engines, being technically complex and their efficiency not credited, had still not found general favour. He was likely to have been unemployed, had not John Taylor engaged him as engineer at Consols. With this appointment, the articles in the *Philosophical Magazine* ceased.

Competition was fierce, and engineer strove against engineer to get the best results. Woolf built two 90-inch engines for Consols and another for Taylor's United Mines, but they did not achieve as high a duty as he had hoped. They were consistently beaten in the ratings and the trouble boiled over when Samuel Grose's 80-inch at Wheal Towan reached the highest duty yet reported: 62.2m in July 1827 and 61.7 in the August. In that month, the best of Woolf's engines, the 58-inch Pearce's at Consols, could only do 43m, equal third, and of his three 90-inch, one was doing 36m and the other two 37m each.

During that summer, Taylor's 90-inch at Wheal Alfred was moved to Consols, rebuilt, and named "Woolf's". The best it had done at Alfred was 44m, but strange things began to happen at Consols. In October 1827, it apparently achieved 63.7m, and went on to do an unprecedented 67m in the November. These spectacular figures caused a furore and a public trial was demanded. This was held on 17 December 1827 and produced a duty of 63.6m.

At Wheal Towan, Grose returned 87m in April 1828, following an annual average of 77.3m. In retaliation for the doubts cast on the Consols engine, Captain William Francis of that mine asked for a public trial of the Towan engine. This took place on 7 and 8 May 1828, and the duty was recorded as 87.2m.

John Taylor, explaining the difference between a trial and ordinary running, wrote

> Now, as the performance of any engine under experiment will be free from those stops or hindrances which we know diminish duty, and which cannot be avoided in the work of a month, we may expect that what would be found in the one case would exceed that of the other; and so it proves to be[37].

In other words, it was to be expected that in the optimum conditions of a trial, the duty would exceed that of normal performance. At his trial, Grose had equalled his best result, but at Woolf's the engine could not even achieve the duty given in the Reporter. Something was badly amiss. The perhaps inescapable conclusion was that Woolf's duty figures were being falsified and that Thomas Lean was conniving in the dishonesty.

This is explicitly stated by John Lean who refused to have anything to do with the September 1827 Reporter and never again worked with Thomas. He started his own Reporter, recruiting mines (such as Wheal Towan) who did not wish to be associated with his brother. In defence of his actions, he wrote that there was no longer a scrupulous regard for the truth, and that there had been intimidation 'by the menaces of self-made men'.

The 14 mines that immediately left Thomas were:

Mine	*engineer:*	Engineer
Balnoon	*engineer:*	Eustice
Crinnis		Webb
Dolcoath		Jeffree
Great St. George		Michell
Perran		Sims & son
Polladras		Sims & Richards
Reeth		Eustice
Rose		Sims & son
St. Ives		Woolf
Stray Park		Jeffree
Towan		Grose
United Hills		Michell
Vor		Sims & Richards
Wellington		Mayne

The breakup of the Lean partnership and John's new Reporter caused a furore. Both Woolf and Taylor were furious: Woolf because his credibility had been attacked, and Taylor because his engineer was the cause of a public scandal and this reflected badly on himself.

JOHN LEAN'S REPORTER, 1827-1831

John Lean (1791-1851) produced his own Engine Reporter between September 1827 and February 1831. Before doing so, he had been co-compiler with his older brother Thomas ever since 1812. The break-up was sparked by the events that took place during the summer of 1827, but these may well have been the culmination of a longer period of unhappiness.

The August 1827 Reporter was the last to carry both their signatures. It did not contain details of Woolf's engine at Consols. The next (September) Reporter was only signed by Thomas Lean and gave details of just 26 engines compared with 38 in the previous month. It had a brief, incomplete entry for Woolf's engine, giving no figures of its performance and no statement of duty. Thomas continued to issue a depleted Reporter until April 1831.

Thomas' September Reporter was dated 10 October. John's first Reporter was dated 11 October, and carried information on 19 of the engines that had previously appeared in the joint publication. That is to say that, within a month, between one issue of the Reporter and the next, nearly 40 per cent of the mines listed had transferred their allegiance to John, perhaps trusting him to provide truthful information because he was known to be a Quaker.

Woolf was by now in serious financial trouble. He had mismanaged his affairs and was forced to mortgage his house in Camborne: even his gold watch had been sold. He was being challenged by young upcoming engineers, and the income from his patents was diminishing. His designs had never been greatly popular, and it was essential to him that the so-called Woolf's Engine at Consols should be seen to be a success. He was desperate to re-establish his position and Thomas Lean was bullied into recording a less than honest duty figure.

This could have been done by tampering with the counter whose box was secured by a lock, devised by his London friend, Joseph Bramah. It is not impossible that Woolf had (illegal) duplicate keys to at least some of the counter boxes that he supplied to Lean, and had fitted one of these to his new engine at Consols. He could then advance the dials by hand. Or, he could have taken the reading from the duplicate counter at Consols, for which as engineer he legitimately had a key. Or, he could have played tricks with the engine, running it on a short stroke at night or in the weekend. Any of these or other methods could have been used at the mines where Woolf had charge of one of his own engines. It is also likely that he, rather than Lean, was calculating the duty of the engines at Consols and United, since, at the time when he visited those mines, Lean's daily workload was considerably heavier than the norm, suggesting that he had assistance.

John Lean was obviously aware that all was not well, and that his brother was being compliant. It was what was happening at Consols that made him break with Thomas. By way of explanation, his second (October 1827) Reporter included a long statement, headed FACTS IMPORTANT TO BE KNOWN.

As few, probably, even of those into whose hands the engine reporters have regularly fallen, have adequate conceptions of the striking improvements which have been made in engines in the last sixteen years, it may not be inadvisable to state that my father began to make his observations in the year 1811. Among the first of the engines which fell under his inspection were those at Dolcoath. One of these accomplished in the first month *13 million* and the other two 8 million each. (I am now speaking of the engines used in drawing water out of the mine.) The last of these has ever since risen to 30 nearly, the other to 35, and the third to 45! In the year 1811, the amount of bills for coal consumed in the mine was £11,179 15s 10d; in the year 1823, it was only £4,592 10s 11d; difference £6,587 4s 11d! The price of coal in 1811 was rather greater than that of 1823; but to overbalance this very considerably, I have to mention that in the year 1811 the adventurers had only six steam whims at work, whereas in 1823 they had 8 besides a steam-stamping mill, which of itself consumed from five to six hundred, or perhaps I may say seven hundred bushels of coal monthly. But the loss which was sustained, independently of this, was incalculable; for such a wretched state were things reduced by *inattention* and *carelessness* of engineers and engine-men, that not a winter passed by without leaving a very considerable portion of the mine deluged.

Now there are a few of the many happy facts which might be clearly demonstrated to have resulted from the publicity which has been given to the duty of steam engines. Numerous others might be stated, if there yet remained an individual so prejudiced as to question the utility of such monthly exposures.

Engine Reporters have been rendered a blessing to the community. They have excited a spirit of enquiry among all those concerned in mining speculations; and among engineers and engine-men, especially, they have been the means of raising and maintaining a commendable emulation; and of stimulating to increased exertions, diligence and attention. Nor will they ever cease to be of *vital* importance so long as they continue to be made with *a scrupulous regard to the truth* and equity as long as he to whom the office of engine reporter has been entrusted, is not to be intimidated by the menaces of self-made men, or shaken from that inflexible integrity which he should steadfastly hold as the *dearest*, the *brightest* gem of his life.

His belief in the importance of the Reporter shines through this, as does his disgust at the pressures that had been put on himself and his brother: pressure to which Thomas had succumbed and which John was not prepared to tolerate. There is no doubt that it was Arthur Woolf that had caused the trouble.

There are no known portraits of Woolf, and we are dependent upon his contemporaries for a description. Samuel Grose said that he was 'a roughish man', and Trevithick called him 'a shabby fellow', referring to him and his friends as 'beasts'. From all accounts, he was brusque, jealous and a poor collaborator: a misanthropist with a consistently bad temper that made him generally disliked. Matthew Loam, who knew him well, described him as 'the pleasantest of fellows when in good humour, but when vexed, which little would do, he was Woolf in nature as in name.' Loam also said that he was 'a fickle friend', a characteristic that he attributed to Woolf's tendency to believe 'any tale which a malicious person did advance' and, although he had a lively wit, 'his hits were keen and stinging'[38].

He became a man of substance in Cornwall, until, through financial ineptitude, he brought himself near to bankruptcy, but his friend John Farey revealed that he held a deep sense of grievance and ill-humour at receiving less money than he thought his talents merited[39]. He saw himself as another Watt, expected to make a similar fortune, and was bitterly disappointed that the world did not give him the rich living he thought he was owed. His disappointment led him to dishonesty and the falsification of records.

John Lean's Reporter was headed 'Work accomplished by the

following steam engines'; this distinguished it from his brother's whose title was 'Work performed...'. The two were similar in layout and the dates of inspection continued without a break, although, due to the haste with which John's first Reporter was issued, two of his customers (Balnoon and the Perran Mines) had not been visited. Details were given of 19 engines at 14 sites.

Another less obvious change that he introduced was in the way that the date was expressed. John being a Quaker (unlike Thomas) regarded the conventional names of the months (January, etc.) as pagan, and this had caused friction between the brothers in the past. In the joint Reporters, the normal month name was given on the masthead and in the dates of visits to the mines, but there was no publication date on its final page. This changed when the Reporters split. Thomas gave 'October 10th, 1827' on the last sheet of his September 1827 issue. John's instructions confused the printer who gave his compilation date as '10th month 11th, 1827' but got it right in the next issue: '8th of the 11th month, 1827'. This terminology continued to be used on the last page of all John's Reporters, but with the generally accepted month name on the masthead and in the text.

The two Reporters continued their separate ways until early in April 1831 when John's ended almost as abruptly as it had begun. By then, all his mines except six had returned to Thomas. He had made visits of inspection to a few of his mines on 29 March and had sent some data to the printer so that he could begin to prepare the text that was to be issued on 5 April, but the work was never completed. There is a printer's proof containing the duty figures for Stray Park, Reeth, and one of the two engines at Crinnis, together with the whims at Polladras and Vor, but it was never issued as a final document.

Things must have happened very quickly. In March 1831 John Taylor added several pages of new material to his paper "On the duty of steam engines" (published in 1829) and sent it to the editor of *English's Quarterly Mining Review*. In this new text he wrote that statistics were 'taken and completed by Messrs Thomas and John Lean, who publish separate lists of the engines that they respectively have charge of'[40]. He finalised his text on Tuesday 22 March, and, exactly a week later, on Tuesday 29 March, John Lean made his last inspections, and a week later again, on Tuesday 5 April, failed to publish his Reporter. That was the date (i.e. the 5th of the 4th month) on the final page of the printer's proof that was not published.

Work accomplished by the following STEAM-ENGINES, in October, 1827.

MINES.	ENGINE, and the diameter of the cylinder.	Load per inch, on the piston.	Length of the stroke in the cylinder.	No. of the lifts.	Depth.	Diameter of the pump.	Time.	Consumption of coal, in bushels.	Number of strokes.	Length of the stroke, in the pump.	Load, in pounds.	Pounds lifted one foot high, by consuming a bushel of coal.	Number of strokes per minute.	REMARKS, AND ENGINEERS' NAMES.
DOLCOATH	76 inches, single.		8 9				Oct. 28th.							
STRAY PARK	64 inches, single.	7,30	7 0				ditto							
WHEAL WELLINGTON	78 inches single.	15,78	6 6				ditto							
WHEAL VOR	Pearce's engine, 68 inches, double.	10,5	7 3				Oct. 3rd, to Nov. 1st.							
DITTO	Western engine, 53 inches, single.		9 0				ditto							
DITTO	Old, engine, 48 inches, single.	7,0	7 0				ditto							
DITTO	Trelawny's engine, 80 inches, single.	14,15	10 0				ditto							
DITTO	Carlcon engine, 45 inches, single.		0 0				ditto							
POLADRAS DOWNS	70 inches, single.	7,5	10 0				ditto							
GREAT	48 inches, single.		9 0				Oct. 17th, to Nov. 1st.							
W.H. REETH	76 inches, single.	15,2	7 6				Oct. 2nd, to Oct. 30th.							
CONSOLS	80 inches, single.						ditto							
BALNOON	50 inches, single.		0 0				ditto							
WHEAL HARMONY	New engine, 70 inches, single.	6,1	9 3				Oct. 4th, to Oct. 31st.							
W.H. TOWAN	Wilson's engine, 80 inches, single.		10 0				ditto							
DITTO	Druce's engine, 80 inches, single.		10 0				ditto							
UNITED HILLS	58 inches, single.	4,84	8 3				ditto							
GREAT ST. GEORGE	60 inches, single.	9,4	10 4				Oct. 5th.							
PERRAN MINES	58 inches, single.		6 0				Oct. 1st, to							
CRINNIS CONSOLIDATED MINES	63 inches, single.	17,5	8 3				Nov. 2nd.							

5. Part of John Lean's Reporter, November 1827, as sent to the printer

35

John Lean's last Reporter covered March 1831. According to the proofs, he had visited Stray Park and Wheel Reeth on 29 March, and had completed his calculations of duty at both. He had apparently then gone to Crinnis Consols, where he had measured the duty of Walker's engine, but had not checked the 53-inch there. These three completed entries gave the interval between his inspections as being March 4 to March 29. Yet Thomas' March 1831 Reporter (compilation date, April 11) also included Walker's engine (but not the 53-inch) at Crinnis, which he had apparently checked for the period March 12 - April 5. It appears from this that John's health had been failing throughout March and that his brother had tried to help him out.

Thomas' April 1831 Reporter (completion date, May 12) was a composite. The first two pages were headed 'Work performed by the following steam engines', as was his normal style, and contained details of his own engines. The third page had John's heading 'Work accomplished by the following steam engines', and contained John's engines with full entries for Vor and Polladras, covering March 31 - May 9, and for Towan (April 4 - May 12), but incomplete listings for Stray Park, Reeth, Balnoon and Crinnis Consols: none of which had apparently been visited. Thomas made no mention of his brother, nor of his Reporter, nor any explanation of why the additional material was being included. Nor was any reference made in any future Reporters.

The only conclusion that can be drawn is that John suddenly fell seriously ill. He died in Camborne in 1851.

It is surprising that so little has been written about John Lean although the country must have been rife with gossip at the time. Thomas Coulson heard the rumours when he was visiting Cornwall in 1829, but hesitated to repeat them in an open letter to Simon Goodrich at the Navy Board[25]. Woolf's friend John Farey did not refer to the situation in his book, nor did William Pole writing twenty years later when the furore had died down. Modern authors do not seem to have noticed that anything was amiss. D. B. Barton even went so far as to say that at this time

> Although [Thomas] Lean included only a fraction of the engines in the county ... these usually included the best then at work and the Engine Reporter was held not only in high but in unqualified esteem[24, p.48].

Nothing was less true.

THE INFLUENCE OF JOHN TAYLOR

It seems very likely that Woolf owned the Reporter from the beginning until about 1829. If this were not so, how else can one explain his arrogance towards the Leans and his belief that he had the right to dictate the duty figures? Without such monetary power, he would have had no authority over Thomas Lean, who danced to Woolf's music more often than might be inspired by mere friendship. Why should he have done so unless he were a paid employee?

Things changed when Woolf left Taylor's mines in May 1830, got into more financial difficulties and retired to Guernsey, where he died in October 1837. The Reporters now seem to have been financed by John Taylor. The evidence for this is twofold. For one thing the Reporters started to contain new tables that illustrated different ways of assessing engine performance: these were all based on experiments being carried out at Taylor's mines. The tables begin in 1833. The second indicator of his interest is a remarkable burst of writing by him, defending the integrity of the Reports. Five separate articles were published between 1829 and 1835; the campaign ended in 1839 with the commissioning of the *Historical Statement*. This degree of personal involvement by such an important and busy man suggests that there was more interest in the subject than pure altruism.

The first of the new tables appeared in January 1833, and showed the 'Average quantities of water per minute drawn from the mines' and was published monthly until July 1836. It was replaced by an annual summary, the first four of which were reprinted in the *Report of the Royal Cornwall Polytechnic Society*, and the 1838/9 figures in the *Mining Journal*[41]. The highest figures always came from Taylor's mines, far exceeding the amounts drawn from the 40-odd other sites, and demonstrating the efficiency of his engines.

Another table, which appeared only once, in February 1834, calculated the number and cost of the horses that would be needed to do the work of several steam engines. Another short-lived table showed the result of attaching a water meter to an engine to measure the amount of steam generated by a bushel of coal. This data appeared in the Engine Reporters of February to June 1836, with the first tabulation reprinted in the R.C.P.S. *Report*[42]. The experiment continued for the remainder of the year but the monthly figures were not published. The December 1836 Engine Reporter

6. John Taylor c.1826 painted by Sir Thomas Lawrence

summarised the annual results (published and unpublished).

It would have been Taylor himself who inspired the experiments on which these tables were based, and arranged for their publication. In doing this, he was responding to criticism, mostly from outside Cornwall, from those who did not understand duty and thought it an inaccurate measure of engine performance. The data that he provided was discussed by the Royal Cornwall Polytechnic Society and in the wider forum of the new Institution of Civil Engineers of which he was an early member.

In 1831, probably prompted by Taylor who was careful about such things, Lean investigated the accuracy with which coal was measured at the mines. He found considerable variation. A parallel survey by Captain William Henwood discovered even greater disparities[43]. This was disquieting because the bushel was about to be abolished as the legal cubic measure for coal[44]. The Cornish answer to this was to manufacture two special coal barrows. One held 188 lbs, or two bushels; the other took two cwt. said to equal 24 bushels[45]. The inaccuracies caused by these continued until, from July 1856, the duty calculations were based on the 112 lbs hundredweight. In the meantime, Lean did not seem to be worried.

Towards the end of 1834, again possibly at Taylor's instigation, he made a list of all the engines then at work in Cornwall. The schedule was included in his *Historical Statement*[46], and showed 104 pumping engines, 14 stamps and 66 whims. John Farey, however, made the first total 111, by including 7 in Tonkin's Reporter[47] which had defected to him from John Lean and thus from Thomas' Reporter. The omission of them was therefore quite deliberate.

At the end of 1834, of the 111 pumping engines, Lean was reporting 55 and Tonkin 19: a total of 74, with another 37, exactly half as many again, not listed. To put it another way, two-thirds of the engines were included in the two Reporters, and one-third was not. Four years later, Captain Henwood identified 152 pumping engines working at the end of 1838; 12 of these were large (between 70- and 85-inch) and were among those not listed by Lean[48]. At that time, he was including 72 pumping engines: a little over half the known quantity. In 1834, his Reporter was also listing half: 55 out of 111.

The defence of the Reporters

While all this was going on, Taylor was vociferously and repeatedly defending the Reporters.

They were in danger of being permanently damaged by the events of 1827 and, now that he had a personal interest in them, he hurried to reassure the public. But there was more to it than that. The scandal had involved one of his engines at one of his mines, and centred on his engineer. Furthermore, he was known to have befriended Woolf and, when he retired, had arranged for a pension to be paid to him by the adventurers at the Consolidated Mines, where the trouble had taken place. Taylor himself could have been seen as condoning the roguery, if not approving it. His own reputation, as well as that of the Reporters, was in jeopardy.

Again and again, he declared that the Reporters were to be trusted. He wrote a lengthy article to this effect in 1828, (almost immediately following the dispute) which was published in 1829 and reprinted, with variations, four more times in the next seven years: the reiteration proving the gravity with which he viewed the situation. The first publication was in his *Record of Mining*; the whole text was repeated and considerably extended in *English's Quarterly Mining Review* in 1831; it was summarised in the *Philosophical Magazine* in 1830 and again in 1836; it also formed the basis of a lecture given to the British Association in 1835[49]. All read like an exercise in damage limitation, and show the Taylor news management machine at work.

It is noticeable that, when referring to what happened in 1827, he wrote that the engine at Consols 'was reported' as doing 67 million, and that the public trial had been certified as being accurate: both carefully worded truths. There were several references to the excellence of the engine at Wheal Towan and the accuracy of the collection of data from it by John Lean. He described the independent, parallel, records maintained at his own mines, which 'generally' coincided with Thomas' Reporter. Again and again, with deliberately chosen phrases, he sought to restore confidence but the vehemence of his protests showed the extent of his own and the public's concern.

Taylor was one of the founder (1832) members of the British Association for the Improvement (later, Advancement) of Science, and was immediately appointed Treasurer. He seized the opportunity of using the status of the Association as part of his campaign, and applied to its Committee of Recommendations for a grant to be used for a project relating to steam engines in Cornwall. Not surprisingly, his application was successful and two payments of £50 were made to him in 1837/8[50]. He paid the £100 to Thomas Lean and his brother Joel as a fee for writing the *Historical Statement*[51]. On completion, it was 'respectfully presented' to the

HISTORICAL STATEMENT

OF THE

IMPROVEMENTS MADE IN THE DUTY PERFORMED

BY THE

STEAM ENGINES

IN

CORNWALL,

FROM THE COMMENCEMENT OF THE PUBLICATION OF THE

Monthly Reports.

COMPILED AT THE REQUEST OF THE BRITISH ASSOCIATION FOR THE IMPROVEMENT OF SCIENCE.

BY THOMAS LEAN AND BROTHER,

REGISTRARS AND REPORTERS OF THE DUTY OF STEAM ENGINES.

LONDON:

SIMPKIN, MARSHALL, AND Co., STATIONERS' HALL COURT.

1839.

7. Title page of 'Historical Statement'

Association's president and Council, and was said to have been 'compiled at the request of the British Association', another play on words. The authors were 'Thomas Lean and brother', who described themselves as Registrars and Reporters of the Duty of Steam Engines. Taylor's name does not appear. The book purports to be the brainchild of the British Association.

The text begins with an extensive quotation from Taylor's paper "On the duty of steam engines" which contained the first of his many assertions of the truthfulness of the reporting procedures[52]. As further evidence of this probity, the Leans then reprinted the official statements on the public trials of 1827 and 1828 on the engines at Consols and at Wheal Towan: the honesty of which trials had never been in doubt, unlike the reporting earlier in 1827. They continued by describing the development of engines over the last quarter century, giving a month-by-month summary of the Reports of 1814, 1821, 1828 and 1835. The concluding analysis mentioned the work of men like Trevithick and Hornblower as pioneers, but repeatedly stressed the 'considerable success' of Woolf, saying that 'To him we owe the establishment, if not the introduction, of the use of high pressure steam, with expansion and condensation'. His engines 'excelled in correctness of construction'. Any remaining rumours of his connection with the Reporters was quashed by the long and circumstantial story about the Daveys.

'And brother'
Thomas Lean was the obvious choice to write the *Historical Statement*. Not only had he the facts readily available, it was appropriate that he should make reparation for the damage done by himself and Woolf. But how did he find time for the writing amid the hectic routines of the Reporters? It was providential that his older brother Joel (1779 - 1856) should have come back to Cornwall just at the right moment to take over some of the work on the reports. Helpfully also, Joel who was an engineer contributed an (unsigned) technical introduction to the book explaining the expansive power of steam.

Joel's arrival was no accident. It had all been arranged. The British Association archives suggest that Taylor applied for his grant in late 1837 or early 1838. Joel's convenient return before this, in the summer of 1837, suggests that he and Taylor had come to an arrangement. By being in Cornwall, Joel was able to look after the Reporters and allow Thomas to be the principal author of the *Statement*. But he did Taylor a far greater favour.

Midway through the compilation of the book, the Reporters changed hands. From summer 1838 they belonged to Joel. This had several advantages for Taylor. First, when the *Statement* was published in 1839, he could truthfully deny that he had any influence over the text or connection with the duty figures; second, it successfully concluded his campaign to defend the Reporters; and finally, Joel could control Thomas. But this concordat had been reached behind closed doors.

"LEAN'S ENGINE REPORTER"

The May 1838 issue of the Reporter appeared as normal, but suddenly there was a change. The June number proclaimed itself to be *"Lean's Engine Reporter"*. For the very first time the family name appeared on the masthead, prominently displayed in gothic-style type. The publication had been known as *The Monthly Duty Paper* because it did not have a proper title, merely the heading "Work performed by the following steam engines". Now, it was Lean's and it was the Engine Reporter.

Unmistakably, the influence was that of the new owner, Joel Lean the younger, second son of old Joel. He had worked for the Fox family at Falmouth, and had taken out a patent in 1812 with Robert Were Fox the younger for 'Improvements in steam engines'[53]. The next that is heard of him is another patent, dated 1825, protecting his invention of 'A machine for effecting an alternating motion between bodies revolving about a common centre or axis of motion; also certain additional machinery or apparatus for applying the same to mechanical purposes'[54]. In this, he is described as a gentleman of Fish Pond House, Bristol. In the summer of 1837, he returned to Cornwall where he stayed for ten years until he went to Neath in South Wales, where he was one of the founders of the Briton Ferry Ironworks, in company with Joseph and George Davey[55].

He worked with Thomas on the Reporters dated July 1837 to February 1847. He was obviously unwilling to be seen to be associating with him, knowing his past reputation, and the signatures on the monthly reports, as on the *Historical Statement*, was always 'Thomas Lean and brother'. He took the Reporters over from John Taylor, and their new title indicated that a new era had begun. The bad old days were over. The Leans were now accountable and respectable.

When Joel returned, Thomas was 53 and had been responsible for calculating duty since 1812. Life had been busy, if not tempestuous at times, and he probably welcomed Joel's enthusiasm for what was for him a new project. There were several innovations, other than the title. The size changed from folio to quarto. The columns were re-arranged into a more logical order. There were alterations to the headings of the tables. For the first time, the weight of bushel was given: 94 lbs. The dimensions of each boiler were cited. The whole appearance of the Reporters was improved and made more professional.

The title change was not merely to *Lean's Engine Reporter*, but to

Lean's Engine Reporter and Advertiser. The new function was explained in a note in the June 1838 issue.

> Considering the extensive circulation which the Engine Reporter has obtained, not only among the mines and those concerned with mining in the county, but throughout the nation, it has been the opinion of many that it might, with advantage, be made the medium of conveying information (whether by advertisement or otherwise), respecting matters affecting the mining interest, and particularly the purchase and sale of steam engines: we have therefore procured the Newspaper Stamp to be affixed, which will greatly facilitate and extend the circulation; and take this opportunity of soliciting orders for advertisements which will be carefully attended to on the usual terms. Any communication addressed (if by letter, post paid) to the printer, Llewelyn Newton, Camborne, will be thankfully received, and obtain prompt attention.

The previously heavy stamp duty on newspapers had been reduced to 1d. in 1836, and in return for this, the Government delivered newspapers free of charge. Perhaps the Leans thought that giving the Reporters the status of a newspaper was likely to extend their circulation outside the county; they possibly also hoped to promote their own columns, rather than those of the new London-based *Mining Journal* (began 1835) as the focus for the sale of engines. Their ideas of turning the Reporter into a mining newspaper never materialised, but it continued to be *Lean's* (or *Leans'*) *Engine Reporter and Advertiser* until it ceased publication early in the next century.

The June 1838 issue had an advertisement for 'An excellent 22-inch cylinder double *STEAM ENGINE*, with working gear, etc.' The next number, July, offered a 24-inch engine, and announced the sale of mining shares by Mr Kirkman of London. Later in the year, there were advertisements for Blundell's Patent Oil for second-hand steam engines, and for Bickford's Patent Safety Fuze. The opportunity was also taken to promote the Leans' own *Historical Statement*, first announced in October 1838, with a column and a half reprint of the review in the *Athenaeum*, which described it as an important work, 'likely to exercise an extensive and highly beneficial influence on general engineering practice'[56]. Taylor may well have been the author of this unsigned review, which extolled the Reporters. (The journal regularly noted the activities of the British Association and the other learned bodies.) The December Reporter announced that the *Statement* had just been published 'in Royal 8vo, price 10s. cloth ... at the request of the British Association for the Improvement of Science'. Publicity for the book continued to appear in the Report until the end of 1840, after which the advertisements gradually petered out.

Another significant innovation was the sequential numbering of each issue, beginning with that of June 1839 which was said to be No. 323, thus making the first publication January 1812. Yet another new feature, as from October 1838, was the printing of a summary in each monthly Reporter. This gave the number of pumping engines included, their total coal consumption, the millions of tons of water lifted and the average duty. These figures were reprinted in the *Mining Journal*.

Throughout this period there was a steady increase in duty, but the satisfaction with this was shattered in July 1835, when the 80-inch engine at Fowey Consols, built by the comparatively unknown William West, did 90 million, followed by 97.8 in the September. A 24-hour trial was arranged in the following month, with the hope of allaying public suspicion, and the engine achieved 125m. Then, as often happened, the performance fell away and the engine tended to be about fifth in the duty list. It appeared in the Engine Reporters until December 1845, but its high performance reawakened arguments about the accuracy of the counters.

It was suggested at the 1844 meeting of the Cornish Miners' Society that it would be better to calculate duty on a stroke measured by a plegometer, rather than by the existing method of ascertaining the number of strokes by using a counter. Thomas Lean attended the meeting and was able to remind the delegates that such an instrument had been tried by John Taylor at his United Mines in the autumn of 1839, and found to be impractical. To recall the experiments to his readers, he reprinted its results in December 1844 and again in January 1845. The average duty as shown by the plegometer was 60.2m, and as shown by the counter was 61.7m: a difference of about 2fi per cent over four months. Lean commented

> The plegomoter used in this instance was complicated in its construction; and its uncertainty became apparent after working for four months; at the end of the year it was laid aside.

Nothing further seems to have been heard of the plegometer.

It was not only the Cornish that were interested in their own technology. Among the visitors from outside the county was the distinguished civil engineer, William Pole, who subsequently wrote a textbook on the engines. He made the curious comment

> These [Engine Reporters] were, and still are, made for a purpose purely local: they were never published with a view to their circulation out of the county, and no efforts have been made to their general distribution; and, indeed, had it not been for the attention drawn to them by parties having no particular concern in them, they would probably, to this day, have been scarcely known of the county[57].

This seems unnecessarily dismissive. The *Historical Statement* had been published less then three years before he was writing, and had provoked a great deal of interest in the scientific community outside Cornwall, making them aware of the Reporters. But long before this, copies of its monthly issues had been widely distributed. The Neath Abbey ironworks, building engines for Woolf, probably received every copy[58], as may their competitors. In 1825, the Institution of Civil Engineers, through its President, Thomas Telford, was asking Davies Gilbert about the method used by Lean to calculate duty[59]. The engineer, William Armstrong, was reading the Reporter by 1839[60]. The ongoing debate as to the respective merits of horsepower and duty in assessing performance was fuelled by receipt of the Cornish publication in the Midlands and the North-East. For many years, copies had been reaching the informed and the influential men of the day.

Thomas Lean died on 1 June 1847, aged 64. He had married Jane Trevenen of Crowan in March 1805, and together they had 17 surviving children, of whom two were girls. They lived at Crowan until December 1829, when they moved to Marazion. Within three years, he was elected a burgess of the town, and in the 1841 census described himself as a merchant, living at Marazion House, a very substantial property standing in its own grounds. In the 1840 tithe apportionment, Lean is listed as owning a number of properties and some land in the area. The sonorous phrases of his Will refer to his 'Freehold and Copyhold and Leasehold Messuages Lands Tenements and Hereditaments' giving evidence of his investments. He left effects worth £1,400.

CHANGE AND DECAY

Thomas Lean II

The partnership of Thomas and Joel Lean ended in the spring of 1847. Thomas died on 1st June, and Joel went off to Neath. Responsibility for the Reporter passed to Thomas' 15th son, also named Thomas (1827-1908), who took over the March 1847 issue. From that date, having been *Leans' Engine Reporter* since March 1844, it reverted to being *Lean's Engine Reporter.*

Within four months he had made a significant change. He added a new column to the tables, giving the duty achieved by burning a hundredweight (112 lbs) of coal. The figure was additional to the existing result of burning a bushel (94 lbs) of coal. Neither his father nor his uncle had taken that step, but it was overdue. William Tonkin's Reporter had been in competition with the Leans since 1835 and used the cwt. William Brown's new *Cornish Engine Reporter* had begun publication in March 1847, and that too used the cwt. Lean had little option. He published the parallel bushel and hundredweight columns until July 1856, when he abandoned the bushel after Joel's death in the February.

Six years after taking over the Reporter, Lean decided to print as well as compile it and the May 1853 issue declared that it was "Printed and published by Thomas Lean (the proprietor) at his Residence and Printing Office at Praze in the parish of Crowan in the County of Cornwall." The claim to ownership was crossed through in pen on the copy in the Cornish Studies Library, and did not reappear. The evidence of this, plus the retention of the bushel during Joel's lifetime, leads me to believe that Joel continued to own the Reporter despite leaving Cornwall. He probably bequeathed it to Thomas who was then free to make changes.

In November 1862, he moved from Praze-an-Beeble to Marazion, and began to issue the Reporter from his printing works in The Square, continuing until he gave up in about 1896.

When Joel was dead, Thomas started a meteorological table in the Reporters, showing the rainfall; this was relevant since Cornwall is a very wet county and the amount of precipitation had a direct effect on the work performed by the engines draining the mines. The data was collected at Praze, at Helston, and at the Royal Institution of Cornwall, Truro, and was logged by the Rain Inspector, M.P. Moyle, esquire, at Lean's Engine Reporter Office. The table appeared from August 1856 until August 1862. It gave, for the current and preceding year, month by month, the number of

8. Thomas Lean II

days on which rain fell, and the maximum in one day.

There was disquiet about the competition between Lean's Reporter and William Browne's new *Cornish Engine Reporter.* William Sims suggested that the publications should be combined, and two Registrars of Duty employed, one working east of Truro (where Browne was most strongly represented) and the other west of Truro (in Lean's country). They should be paid for their work, and every quarter would change areas. His ideas were considered by the 1857 meeting of the Royal Cornwall Polytechnic Society, whose members felt that efforts should be made to induce Lean and Browne to unite and work jointly[61]. Lean acceded but neither Browne nor his co-producer William West seem to have been present, and were unlikely to have agreed had they been there. In the event, there was no change in the situation. Lean was being conciliatory, but it is doubtful whether he, any more than Browne, would have been content to be an employee of the society.

One of the principal criticisms made of Browne was that he made allowances for the steam used to heat the miners' dry, where they hung their wet clothes. Despite his explanation of his methodology, and his repeated insistence that the deductions were all carefully calculated and had general agreement, his calculations were mistrusted by his vociferous enemies. His Reporter ended in April 1858, and, with a timing so exact that it cannot have been accidental, Thomas Lean admitted in July 1858 that he too made certain allowances, which he described as being 'proportionate'. Curiously, this statement has been forgotten by the anti-Browne faction[62].

The Polytechnic Society intervened again. In May 1859, Lean's Reporter (no. 573) carried an imprimatur, displayed prominently in capital letters below the title: Published with the approval and recommendation of the Royal Cornwall Polytechnic Society. This statement was carried on every issue until November 1886. From May 1859 also, another note below the masthead added: Established 1811. This continued throughout the remaining life of the Reporter.

The end

By the time that the R.C.P.S. were ready to declare their allegiance to the one remaining Reporter, the duty war was virtually over and Lean's publication had almost ceased to matter. Everything became irrelevant when disaster hit the whole Cornish mining industry in the 1860s. So far as tin was concerned, the American Civil War caused stagnation in the tin-plate trade, and the development of overseas mining ended the domination of world markets. Copper mining was thrown into crisis when places such as Cuba, Chile, South Australia and Michigan expanded their production until by 1868 the output of the last of these was greater than that of all the Cornish mines. Thousands of miners emigrated and there was hardship among those who remained.

In the mines that stayed open, the maintenance standards fell disastrously. Pitwork was neglected; there were leaking boilers, broken fire-bars, cisterns full of mud, and corroded pumps. Enginemen were no longer paid duty money, except at Dolcoath where a premium system for saving coal was still in existence in the 1870s. It was symptomatic of the decline in engineers' status that, from the beginning of 1862, their names ceased being given in the Reporters. By 1870, the job was being given either to disabled men or to young lads. Fewer mine owners saw any point in paying a fee to Lean for inclusion in the monthly lists. Some of them compromised by only paying for their better engines and were not interested in the performance of old crocks that were soon to be scrapped.

William Teague deplore this in a speech made in 1882 to the Mining Institute of Cornwall

> In order to estimate the duty fairly, I would suggest that every engine on a mine should be reported and not let any particular one shine at the expense of some old rattletrap, struggling for existence: that is to say, not making exception of the known best to the exclusion of the worst in your calculation; and with this statement of average duty, particulars of all coal consumed should be published[63].

He also referred to the efficiency of Cornish engines having been 'very much exaggerated at one time' but did not name any names.

The apathy was apparent in the columns of the Reporter: the data for one engine, for example, might not include the amount of water drawn; another might have no figure for oil and tallow consumption; a third did not have its horsepower given. A common remark was 'Counter idle'. It scarcely seemed to matter that Lean was absent for four months in 1878. The August issue was produced as usual, but he next to appear was a composite for September - December (finalised on 23 January 1879) which carried a note

> Having visited Pennsylvania since the August Report was issued, on some legal business which necessitated my

attendance there, the present issue embraces all the intervening period.

The Reporter limped on, with a steadily decreasing content. It listed twenty engines in 1870, but by 1890 it was down to thirteen, and fell to seven in 1895. By now, not only the rattletraps were being excluded, even big new engines in important mines were not being shown: for example, a 70-inch at Tincroft that was started in December 1893[64]. Elsewhere in Cornwall, engines were being sold at less than their scrap value, and it was often not worth recovering the pitwork.

By now, Thomas Lean was becoming an old man, and although he was still running the Reporter at the beginning of 1895, he had retired by the spring of 1897: the time of his 70th birthday. (We do not know exactly when he gave up because no copies of the Reporter between these dates seem to have survived.)

For many years, he had been heavily involved with local affairs, in addition to conducting his business. He had moved to Marazion in 1862, and taken the town to his heart. His obituaries record how greatly he was respected, and how many were his public offices. Within two years of settling in Marazion, he was elected Capital Inhabitant, and was Councillor and Alderman from 1864 to 1888. He was mayor eight times (1875-80 and 1883-6), and the first chairman of the Marazion Town Trust between 1885 and 1906, only retiring because of increasing infirmities. (The Trust was one of only two in Cornwall, formed after the Local Government Act of 1894, to own and hold all the old regalia, archives and memorabilia. Lean's appointment shows his deep commitment to the organisation.) On the dissolution of the old Corporation, he was elected to be the first chairman of the new Marazion Parish Council, 1894-5.

He was a benefactor to the town in many ways, and it was through his efforts that a modern water supply was obtained. In 1885, he had built for himself and his family a substantial grey stone house, lying well back from the road, with a large garden. He named it The Gew, and from its top floor observatory there is a wide view of St Michael's Mount and the panorama of the Bay.

A deeply religious man, he was a great friend and supporter of the British and Foreign Bible Society. He was prominently identified with the Wesleyan Methodists (as were many members of the extended Lean family) but in later years he turned to the stricter disciplines of the Plymouth Brethren, joining their meeting at Penzance[65]. Before that, he worshipped at

Nancegollen Methodist church, and the extent of his involvement is shown by the evidence of his name on the foundation stone that was laid in 1859. When he died, on 11 July 1908, in his 82nd year, he was buried in the churchyard there and his grave given a prestigious position, just outside the main door, so that he might not be forgotten.

After Lean retired, William M. Prout of Redruth is said to have been Registrar for a few months[66], but by April 1897, the task was being carried out by James Champion Keast, A.I.Mech.E., lecturer in Practical Mechanics at the Camborne Mining School. The printer was the Cornubian works at Redruth, and the publication continued to be *Lean's Engine Reporter and Advertiser.*

By now, both the mining industry and the Reporter were almost in terminal decline. Although fourteen engines were listed in July 1902, there were full details of only four. The last issue that has survived is that for July 1904, number 2015. This shows fifteen pumping engines, with the duty calculated for four: the 80-inch at Carn Brae & Tincroft Ltd, the 85- and the 65-inch at Dolcoath, and the 80-inch at the Bassett Mines Ltd. According to the footnote, their average duty was 55.2 million. The stamps at Dolcoath and Bassett were working, with a duty of 46.5m at the former, and at the latter 51.2m and 26.4m. The details and the layout are muddled, and many of the remarks are incomplete. As a record, it was scarcely worth publishing. Its compilation date was August 30, 1904.

James Champion Keast (1865-1914), the last of the Registrars, was an authority on mining engineering. Born at Illogan, he worked at the Tuckingmill Foundry from 1880 to 1896, first as an apprentice, latterly in charge of the erecting shops. He enrolled as a pupil at the Camborne Mining School in 1882, began taking classes in 1886, and in 1896 was appointed lecturer in Practical Mechanics, a post he held until his death. He took a deep interest in the formation of the Cornish Institute of Engineers and was one of its first vice-presidents. As secretary of the Trevithick Centenary celebrations at Camborne, he was a member of the committee responsible for erecting the Trevithick statue at The Cross, and the founding of the Trevithick scholarship at the School of Mines. He was elected Associate member of the Institution of Mechanical Engineers in 1901.

In politics, he was a vice-president of the Liberal Association and during elections worked most strenuously for the cause, but his real devotion was to the Wesleyan church, where he was class leader, Sunday school teacher and local preacher. He was an enthusiastic temperance

9. James Champion Keast

worker, often boasting that the Camborne circuit contained more chapels than public houses. At the School of Mines, he was one of the most lucid and effective lecturers that the students ever had, and his services as a local preacher were in great demand, but it was this popularity that led to his early death from pneumonia, contracted by getting wet walking from a preaching engagement in one of the more remote parts of the Camborne circuit[67].

He died in the middle of lecturing to his students at the School of Mines, and those students pulled the hearse to his funeral at Centenary Methodist church in Camborne. Three thousand people were reckoned to have paid tribute to him on that occasion, giving thanks for the life of one of the prominent men of the town, whose cheery enthusiasm and vitality were an inspiration to all[68].

There are very few copies left of the Reporters compiled by Keast, and we can only guess that the last issue was that for July 1904. After his death, the thorough house clearing carried out by his son, Jack, destroyed many documents that lurked in the cupboards, and unfortunately lost many of the Reporters to us[69]. The journal had effectively ended with the retirement of Thomas Lean, but Keast did his best to continue what had now become an almost unwanted and unnecessary record. The sun had set on the greatness of the Cornish beam engines. Their history was bound up with that of the Leans, and there can be few other families who can claim to have presided over the birth, life and death of a major industry. Through the generations, for almost a century, they had recorded these magnificent machines.

For us today the Leans' name is synonymous with Engine Reporters, but there were at least four other series. Two of them reflect the bitterness with which the duty wars were fought, and two were instruments of management control. Tonkin's Reporters appear to be a continuation of John Lean's protest but by the time that he started publishing it was Joel rather than Thomas Lean who was in charge. William Browne, on the other hand, was to a great extent acting as publicist for William West, although it was said to have been the inaccurate calculation of duty that caused the split with Lean. John Taylor's two Reporters were integral parts of his business system, through which he monitored his investments in the mines of Flintshire and Mexico. But whatever the reasons for setting up these different Reporters, their style and presentation were modelled on the pattern established by the Leans.

TONKIN'S ACCOUNT

William Tonkin of Redruth began to issue his *Account of the work performed by the following steam engines* in October 1834[70]. It probably ceased early in 1841. No copies of it have survived to the present day, and were scarce even 40 years after its publication, since the *Bibliotheca Cornubiensis*[71] could cite only one issue: that for May 1838. Even immediately after its demise, it was obviously not well known, and W. J. Henwood, writing for the Royal Geological Society of Cornwall in 1843, felt obliged to explain that it was similar to Lean's Reporter and 'contained the reports of engines not recorded by them'[72].

Farey said[73] that the data given by Tonkin was nearly the same as that of Lean, the principal difference being that, from August 1835, he began using the hundredweight (112 lbs) instead of the bushel: a move not followed by Lean until 1847. (The legal requirement to substitute weight for heaped measure came into effect at the beginning of 1835[74]). The copy quoted in the *Bibliotheca* was apparently printed on a single folio sheet by F. Symonds of Redruth, whose premises were 'two doors below Andrew's hotel'.

Within four months, by January 1835, he was listing 19 engines as against Lean's 55; by December 1840 he had 15 against Lean's 60[75]. As in all the Reporters, mines and engines come and go in a manner that was doubtless logical at the time but is less obvious today. It is clear, however, that some mines that had defected from Thomas Lean to John Lean in 1827 defected again to Tonkin, for a time, but because none of his Reporters has survived, one cannot be sure exactly which they were. John Farey compared the engines listed in Lean's and Tonkin's Reporters of January 1835, and noted that Lean's list of the engines at work at the end of 1834 (in the *Historical Statement)* omitted seven engines. These are likely to have been Tonkin's and can be identified by cross-checking the Lean table, his January 1835 Reporter and his brother John's Reporter[76]. There were the 38-inch at the Perran mine, the 40-inch from Penwith, two (53-inch and 56-inch) from United Hills, the 70-inch at Polladras Downs and the two 80-inch at Wheal Towan. Other hints by Farey indicated that Tonkin seems to have had Godolphin, Balnoon, Perran, Stray Park, and possibly Tresavean. This would mean that he was listing some of the important engines of the time.

Unlike many of the other reporters, Tonkin was an engineer. One of his contracts was to build a new engine for the United Portsmouth, Portsea and Farlington Waterworks Company in 1841[77]. The existing site was in the marshes and was regularly inundated by the sea; Tonkin therefore recommended that the works should be moved further north 'on the hard ground near the turnpike road at the foot of Portsea Hill', but before this could be done, the sea finally destroyed the old engine in August 1841. Tonkin built his new engine on the new site, and presented the company with a bill for £1,400 at a time when they were nearly £10,000 in debt.

Relationships between himself and the Directors seems to have been very unhappy throughout the work on the contract, and became really acrimonious when he presented his bill. It emerged that he was in cahoots with the Secretary of the Company to share the profits of the sale of the new engine, and both of them were accused of attempting additional fraud by issuing bogus share certificates. No legal action was taken, perhaps from lack of sufficient evidence, but the Secretary was dismissed and Tonkin scuttled back to Cornwall[78].

The engine itself seems to have been a success, and the economies resulting from it enabled the Company to be in financial surplus within three years. The *Hampshire Advertiser*, ignoring the scandals, reported in 1841

Mr William Tonkin, engineer from Redruth in Cornwall, has lately substituted a Cornish boiler of his manufacture for the old wagon boiler at the Portsmouth waterworks, by which he has reduced the consumption of coals from 11 tons to 4fi tons per week and supplies the town with water for 4 hours less each day than was previously required[79].

So far as his Reporters were concerned, their beginning in 1834 follows closely upon the mysterious ending of John Lean's series in March 1831. The defection for the second time from Thomas Lean's Reporter of some of the important mines is further evidence (if more were needed) of the distrust felt for Lean, even after Woolf's disappearance from the scene. What is not clear, however, is the extent of Tonkin's own probity in his engine reporters: the Portsmouth story suggests that he too could have been bought.

He stopped compiling them in the spring of 1841 when he began at the waterworks. He is lost to my sight until he appears in the 1866 *Doidge's Directory of Redruth* as an Engine man, living at Tear Waste.

Extensive research has failed to find any further information.

Letters from the Portsmouth Waterworks in 1841 were addressed to him both at Redruth and Penryn, but his name cannot be distinguished in the faint records of the 1841 census. Nor can he be traced in later censuses. Interestingly, however, he was described at the time of his preliminary appraisal at Portsmouth in 1840 as being a London engineer, but the 1841 *Hampshire Advertiser* places him as being from Redruth.

His Engine Reporter began three and a half years after John Lean's ended, and there may well have been a connection between the two events, particularly since in 1825 Blanche Lean (sister to John and Thomas) married a John Tonkin of Redruth, albeit from another branch of the family. One wonders if William Tonkin was persuaded by someone that he should continue the opposition to Thomas Lean but found the job too arduous and less rewarding than he had hoped. The Portsmouth adventure suggests that he might not resist any chance of making money.

BROWNE'S *'CORNISH ENGINE REPORTER'*

William Browne was born in London in 1799, lived most of his life in Cornwall, and died at St Austell on 28 November 1878. His *Cornish Engine Reporter* was published from 1847 to 1858.

As a young man, he worked in the office of various mines and by 1841 he was John Taylor's agent at Polgooth and at Wheal Anna. When the former mine closed in 1842, one among many victims of the falling price of tin, he set up in business as a commission agent, buying and selling mining equipment and trappings, and also acted as auctioneer at St Austell. In March 1847, the first issue of a new Engine Reporter appeared, with his name on the masthead as proprietor.

Although he had a mining background, he was not an engineer and had no obvious qualifications for the job of compiling Reports. His engineering connections came through his marriage to Elizabeth Giddy Grose, the daughter of Samuel Grose (1764-1825), and sister to Samuel Grose junior (1791-1866), both engineers of distinction. Through them, he met William West.

The younger Grose engaged William West (1801-79) as his assistant at Great Wheal Towan in 1827, and, three years later, West became engineer to John Thomas Treffry (born J.T.Austen) (1782-1850) at Fowey Consols. He designed for him an 80-inch engine, costing £2,097, and, never one to be modest, boasted, even before it was erected, that it would 'enable us to vie with any in the County'. His claims were substantiated when it did 125 million in a public 24-hour trial in October 1835. This caused a sceptical furore, in which James Sims, with Hocking and Loam, tried to get the result discounted and the trial repeated over a 3-day period, but West, supported by John Budge who had supervised proceedings, challenged Sims to a competition of their respective engines. The dispute, waged in the columns of the *West Briton*[80], gave West a great deal of publicity, but his engine only recorded 88 million in the following month and, in Lean's Reporter, never again reached 100m.

West's employer, Treffry, was a successful, but irascible, entrepreneur. He had substantial interest in china clay and granite; dealt in iron, timber and coal; carried on businesses in lead smelting, lime burning, and candle making; developed the harbours of Par and Newquay, and built roads and railways. He was known as 'The King of Mid-Cornwall'. Fowey

10. William West of Tredenham, civil engineer

Consols was second only to the great Dolcoath in the richness of its yield, and he was described in 1848 as 'the greatest employer of miners and other labourers in the West of England'[81]. He and John Taylor were bitter rivals.

It was inevitable that there would be trouble when, in 1840, Taylor erected at the United Mines a new 85-inch engine, built by Hocking and Loam, which in September and October 1842 achieved the highest duty ever to be recorded by Lean: 107 million. It continued to dominate the lists, frequently beating its nearest rival, often the engine at Fowey Consols, by a clear ten per cent.

West stood this for three years and then, at the end of December 1845, removed all his engines from Lean's Reporter. Ostensibly, this was the result of his private rivalry with Hocking and Loam over the duty of their respective whim engines. Matters came to a head when Lean claimed to have discovered that West was incorrectly stating the weight of his kibbles, thus exaggerating the work done, in order to get a higher duty. This was the reason given for West's action in withdrawing from Lean, but there are three other possibilities as to why he reacted as he did. Firstly, he and Treffry may have been piqued because the Fowey Consols engine was being beaten by one of John Taylor's. Alternatively, West may have suspected (or known) that the rival 85-inch was not being honestly reported, because its high performance was said to be continuing instead of dropping, as was usual after a few months. Possibly also, he may have been convinced that his own engine was being misrepresented. Whatever the cause, West and Treffry broke with Lean, and none of their engines were reported in 1846.

A year later, in March 1847, the first issue of the *Cornish Engine Reporter* was published. According to the masthead, it was 'Printed for the Proprietor, William Browne, of Charlestown, St Austell, by Jacob Halls Drew, St Austell, Cornwall'.

It contained information on 23 pumping engines at 17 mines, and on the 22 rotary engines at seven of those sites. William West was superintending engineer at 13 of the mines, his brother (John) at three, and Hocking & Loam at one (Drakewalls). The column headings gave, in the main, the same information as Lean's Reporter, with variations of wording and order. There were additional columns recording the actual horsepower employed and, the oil and tallow consumed. Duty was based, not on Lean's bushel of coal, but on the now mandatory hundredweight.

The other significant feature was a column headed 'When made, maker's name and from what Engineer's Drawings'. In that first issue, West had designed 14 of the 23 pumping engines (John West two, Eustice one, six were unknown) and 13 of the 22 rotative (eight unknown, Webb one). The Reporter therefore gave information both on the original designer of the engine and on its current superintendent: on both of whose skills the duty depended. William West had this dual role in 27 out of the total of 45 engines reported.

The charge for inclusion on Browne's Reporter was 7s 6d per month for one engine; 6s per month for two; 5s for three or more. An annual subscription of 12s secured copies for those not connected with the mines. The Leans originally charged one guinea (£1 1s) per month per engine listed[82]. Their annual subscription was £1 in 1825[83].

Both Lean and Browne watched each other. The first to make changes was Thomas Lean who, in May 1847, almost as soon as Browne began his Reporter, started to calculate duty in terms of the 112 lb hundredweight. In September 1847, Browne added a new column, giving 'Consumption per horsepower hour' to match similar information given by Lean. Both men sent summaries to the press, with Browne's being the more elaborate, giving the average performance of rotary and stamps engines, as well as those used for pumping, naming those in the three categories with the highest duty. The data, like Lean's appeared in the *Mining Journal* or the *West Briton* or the *Royal Cornwall Gazette*.

Both the layout of the Reporter and the technical notes which explained its policy appeared to be the work of an engineer, but were invariably signed by Browne. He himself dealt with the routines involved with the publication, and the February 1847 number began with a rather worried introductory Note

> The First issue would have appeared much earlier, but from an unaccountable delay attending the printing. And in presenting it in it present unfinished state, it is necessary to remark that, in many cases, the counters are not fixed; and, in others, some omissions have arisen in giving the required information, etc., etc. But if the parties interested will be kind enough to waive this for the present, a Number or two will put all right.

This was followed the next month with a more positive statement

> The Rules in conjunction with this Report have met with the direct sanction of every Agent and Engineer; and as ACCURACY and TRUTH must mainly depend upon the same

being carried out, I have much pleasure in observing the spirited determination of some, fully to enforce them; and I trust that in a few months it will be general. The advantages arising to the mines have honourably borne testament to by several distinguished Agents.

The next issue noted that at five named mines the steam pipes were left uncovered so that the boiler houses could be used as 'drys'. The matter was obviously causing some debate and it was said, in May 1847, that, although some establishments were hoping to have some allowance made, no action had been taken but the general opinion was that one-tenth would be fair and an announcement would be made in the next issue. As to the measurement of coal

The Barrows have been carefully weighed; and as uniformity is absolutely necessary to establish correctness, I would observe that every Barrow ought to be made so as to contain 2 cwts. of coal heaped without any projecting; and the continuance of the same practice of heaping the coals when they are wet will give about a fair allowance for the proportion of water in them at that time.

The June 1847 issue came out with specific proposals on allowances

The difficulty of fixing what is a fair deduction from the consumption of coals when the Steam Pipes and Boilers are left uncovered for the purpose of drying the men's underground clothes is too great to arrive at with absolute certainty. The escape of caloric is about the same, but from the amount of power employed, and consequent consumption of coals, a considerable variation must exist; and the following is proposed as a standard; subject however to further consideration, and on which as on any other matter concerned with the Reporting any Communication would be esteemed a favour:

Engines working under Two Strokes a minute : One Seventh
ditto Two, & under Three Strokes   : One Eighth
ditto Three, & under Five Strokes   : One Ninth
All others   : One Tenth

(These proposals should be kept in mind when considering the later criticisms made of the Reporter.)

The August 1847 issue announced

The next number of this report will contain some slight alterations, and give the general data on which the calculations are founded. The QUALITY of the coals and the waterway in the clacks are subjects of great importance, and from facts which might be given, it is clear that strict attention to them will ensure an amount of economy which the casual observer is not prepared to admit.

In September, an additional column was included giving hourly consumption of coals per horsepower. The introductory note explained how horsepower was calculated for single and double engines: an explanation which, by the following January, had been extended to include the methodology for combined pumping engines and for rotary engines. The Note was continued by further comment on the fuel.

The variation in the quantity of coal, even from the same pit, is very great; and the economy in attending to this will be found to vary from one-twelfth to one-fifth of the consumption. This presents ample inducement for purchasing the best coals, and thus securing more regular and greater duty from the coals.

The methods of measuring horsepower caused some criticism, and, in January 1848, it was said that the standard adopted had 'the united consent of several eminent Engineers' but 'There are, however, others of the first respectability who differ from this scale, but they have not furnished us with their views'. All of these technical notes were signed 'William Browne'.

It may seem strange that Browne, with no engineering experience, was able both to raise and resolve such critical problems: yet his was the only name on the heading to the Reporter. Significantly, however, he never claimed any responsibility for it. The advertisements for mining goods that appeared in the early issues cited him as Commission Agent, and, having rejoined Polgooth when it opened, he gave his occupation in the 1851 census as Mine Agent and Auctioneer.

The man with the incentive, the engineering knowledge and the money was William West. Just as Woolf had used Thomas Lean, so West used Browne. Both of them needed a front man who was prepared to allow the use of his name for the engineer to shelter behind. There can be little doubt that they were West's Reporter*s*, and those people, even today, that criticise their accuracy should blame West rather than Browne who merely acted as a clerk.

The question remains as to the motives behind these Reporters. Were they self-seeking publicity, with modified regard for the truth, or were they an honest listing of engineering achievements? How far were they a reflection of Treffry's jealousy of Taylor? Or were they a protest against Thomas Lean? The emphasis on the need for accuracy and truth (as stated in the introductory note to the second issue) is very reminiscent of the call for a 'scrupulous regard to the truth' made by John Lean twenty years previously, or was this itself a spurious attempt to claim justification?

Why should West have wanted to be bothered with the Reporter? By the mid-1840s, he was a wealthy man. In addition to his position with Treffry, he was building engines for a great many of the mines in eastern Cornwall and West Devon, and had extended his operations to Ireland, Wales, Spain and Belgium. He had supplied the great engine for the West London waterworks: a contract that led to work for other water and canal companies. In 1848, he established the foundry and engine works at St Blazey, and, a few years later, bought the Lower (main) Foundry at St Austell, thus becoming designer, builder and superintendent of most of the engines in his Reporter.

It may be thought that, with his business interests so well established, he hardly needed the Reporter as a publicity medium. He obviously believed he did. It provided a useful record of his engines, was a continuing demonstration of his technical expertise, and gave evidence of his standing in Cornwall. Besides which, it probably pleased his patron, John Treffry.

Whatever the reasons for its existence, it fails as a record because few copies have survived. Even in 1863, William Morshead could not find a set of them for a talk that he was giving to the Institution of Civil Engineers, nor could he trace a 'condensed abstract' of them[84]. The virtual disappearance of Browne's Reporter has meant that there is no continuous record of the work of William West, one of the most important Cornish engineers of the period.

The Reporter suddenly ended with the April 1858 issue. By that time, West had become a very wealthy and a very busy man, with contracts

CORNISH STEAM ENGINES.—(*From Lean's Engine Reporter, for August,* 1852.

The number of Pumping-Engines reported for this month is 15. They have consumed 865 tons of coal; and lifted 7,000,000 tons of water ten fathoms high. The average duty of the whole is, therefore, 49,000,000 lbs. lifted one foot high by the consumption of a bushel of coal weighing 94 lbs.—The following have exceeded the average duty:—

Mines.	Engines.	Length of stroke in the cylinder.	Length of stroke in the Shaft.	Load in Pounds.	Load per square inch on Piston.	Number of Strokes.	Strokes per Minute.	Coals in Bushels 94 lbs.	Actual horse power employd.	Consumption per horse power per hour.	Mills lifted one foot high by consuming one bushel of coal. 94 lbs.	Ditto consuming 112 lbs.	ENGINEERS NAMES.
Great Work	Leeds 60-inch	9,0 ft.	7,0 ft.	55,343	15,2	288,000	7,1	1824	83	3,0 lbs.	61,1	73	P. Roberts.
W. Wh. Treasury	40-inch....	9,0	8,0	27,961	19,7	264,000	6,5	1197	44	3,7	49,7	59	Sims and Son.
Perran St. George	60 & 100-inch	9,0	8,0	81,187	25,3	231,000	3,7	2696	112	3,3	55,6	66	Sims and Son.
East Wheal Rose.	Penrose's 85-inch ..	10,0	9,0	91,986	14,5	162,000	4,0	2354	100	3,2	56,9	68	Hocking and Loam
Ditto........	Michell's 85-inch ..	10,0	6,8 / 9,0 / 3722	89,510	14,6	106,000	2,6	1512	65	3,1	58,2	69	Ditto.
Ditto........	Purser's 56 inch..	10,33	8,1	57,085	18,1	153,000	3,7	1328	51	3,4	53,2	63	Ditto.

from North and South America, Africa, Australia and New Zealand. The publicity had served its purpose and he had no further need of it. More than 40 engines were now listed in the Reporter (as opposed to only 15 in Lean's) but both West and Browne had outgrown it. J. T. Treffry had died eight years earlier and his rival, John Taylor, had become a man of national importance.

Browne had grown tired of the routine and was anxious to continue his own entrepreneurial activities. His first foray had been into china clay and in 1849 he had taken out a patent on a method of pulverising it[85]. He acquired a lease on the Leeds pit on Tresowes Hill and sent a sample to the 1851 Great Exhibition (together with a copy of his Reporter, unlike Lean) and another to the 1862 Exhibition[86]. From 1856 to 1878, he was in partnership with West on a clay sett previously worked by Treffry at Wheal Anna (Rosvear)[87]. In 1859, he bought land at Brixham, Devon, with a view to mining iron ore and quarrying limestone; soon afterwards he was leasing iron mines on the eastern edge of Dartmoor. In total, from 1849, he was leaseholder or owner of 19 mines and 3 clay pits in Devon, and of 3 and 8 respectively in Cornwall. It is quite obvious that he must have received substantial sums of money from West and, significantly, when Polgooth mine, at which he was agent, closed down in 1857, he allowed West (by way of recompense for past favours) to buy by private contract for £291 all the tin halvans and slimes which were estimated to be worth £1,500[88].

It is difficult to form an unbiased judgement of Browne's Reporter. Most of the contemporary criticism seems to have come from those who felt threatened by it and by its hold on the eastern mines, to say nothing of its inclusion of the massive Devon Great Consols on the farther bank of the river Tamar. West was establishing himself as the chief engineer in that area and had a locally based Engine Reporter to publicise his achievements.

The controversy continued long after the ending of Browne's Reporter. A man who felt particularly bitter was Matthew Loam (1819-1902), son of Michael Loam, (1797-1871) a pupil of Woolf. As late as 1896, Matthew Loam told Nicholas Trestrail that Browne's Reporter was 'a notorious partisan, encouraged and supported by one of the Eastern mines'[89] and Trestrail (1859-1922), briefed by Loam, spoke dismissively about circumstances that had arisen before his own birth, and of which he had only circumstantial (not to say, biased) knowledge. He said that 'the monthly *Cornish Engine* Reporter did not find much favour, and the publisher and the publication long ceased to exist'[90]. He was also critical of the allowances given for unlagged pipes in the Dry at some mines, but chose not to remind his hearers that Browne/West had explained the methodology of their deductions, whereas Lean, who was making similar allowances, did not reveal that he was doing so until Browne's Reporter had ceased and did not explain his calculations.

It is worth noting that Browne's was the third of the Reporters that had been established as a protest against the Leans, and this in a period of only 20 years. First, there was John Lean's, which ran from 1827 to 1831; then William Tonkin's, 1834-1841, and finally, Browne's, 1847-1858. Both John Lean's and Tonkin's were directed against Thomas Lean the elder, whereas Browne's was published during the editorship of the younger Thomas Lean, but its gestation was in the lifetime of his father, and its origin was probably due to his malpractice, even if William West stood to benefit from its publication. The extent of the dissatisfaction is illustrated by the fact of at least one mine (Stray Park) defecting three times from Lean to each of the protest Reporters, as they appeared.

Of these three, Browne's was the most significant, not only because it lasted the longest, but also because it threw the whole reporting system into confusion. The number of engines listed by Lean was never again as great as it had been before Browne's arrival. In 1841, for example, Lean showed the results from 50 engines; in 1858, there were 15; and in 1860, two years after Browne had finished publishing, Lean was only listing 24 engines.

EDDY'S REPORTER

Writing in 1831, John Taylor revealed that

The same [i.e. Cornish] system and the same method of estimating duty have been adopted in the lead mines of Flintshire, and the advantages are sufficiently manifest[91].

The significance of this became clear when I discovered a single copy of what was obviously an Engine Reporter in the Goodrich collection in the Science Museum Library in London. The printed signature at the bottom was Stephen Eddy, and there was reference to the Mold mines. Another issue came to light in the Clwyd Record Office, with the stamp of a Birkenhead painter and decorator on the back. No other copies have so far been found. The dates of the Reporters were 1829 and 1833, and they were obviously being compiled each month for John Taylor during the period he was mining in North-East Wales. It is well known that he introduced Cornish methods of working into his enterprises outside the county, and, although unpopular, substituted the tribute system of employment, but here was proof that he was concerned about the performance of steam engines in an area where coal was plentiful.

Since the Middle Ages, Flintshire had been an important source of lead and zinc but by the 1790s, intensive mining had exhausted the surface deposits and drainage difficulties were hampering extraction at depth. The geological structure of the ore-beds created a problem: there was too much water below the surface, but not enough above ground to drive waterwheels. The only saving grace was the availability of coal for steam engines, but the early designs were not sufficiently powerful and the experiments were expensive failures.

In 1822, John Taylor was appointed mineral agent to Lord Grosvenor, the land owner, and almost immediately installed a 50-inch steam engine at the Halkyn mine, replacing an adit which the men had been driving for the last four years. He then put engines at two other Grosvenor mines: a 63-inch at Milwr and a 50-inch at Gwernymynydd. He next tried to reopen the flooded mine at Llynypandy by sinking a new shaft that he christened 'the Conqueror of Wales', drained by an 80-inch engine with the same hopeful name, but even this massive effort could not cope with floods and the mine was closed two years later in 1829[92].

Work performed by the following STEAM ENGINES, in September, 1829.

ENGINES, and the diameter of the cylinder.	Load per square inch, on the piston.	Length of the stroke in the cylinder.	No. of lifts	Depth.	Diameter of the pump.	Time.	Consumption of coal, in bushels.	Number of strokes.	Length of the stroke in the pump.	Load, in pounds.	Pounds lifted one foot high, by consuming a bushel of coal.	Number of strokes per minute.	REMARKS, AND ENGINEERS' NAMES.
MOLD MINES Upper Pandy, 60 inches, single.	8, 65	10 0	1	20 3 0 13 0 0 31 0 0 12 3 0	22½ 16 10 16	Aug. 31st to Sep. 30th	2795	301530 180800	8 6	44626 6346	11,928,270	7	Drawing all the load perpendicularly. Main beam over the cylinder. One balance bob at the surface. GROSE.
DITTO Taylor's Engine, 60 inches, single.	11	9 0	1	33 4 6 11 0 0 10 4 0 11 0 0 16 3 0	18 13½ 19½ 18½ 8	Ditto	1173	339910	8 0	18850 103	31,191,661	7,66	Drawing perpendicularly 89 fathoms, and on the diagonal 32 fathoms. Main beam over the cylinder. One balance bob at the surface, and one bob underground. BRATT.
DITTO Old Engine, 60 inches, single.	11,35	9 0	1	17 0 0 43 3 6	10 22	Ditto	2703	258970	7 6	46622	33,300,831	6	Drawing all the load perpendicularly. Main beam over the cylinder. BAWDEN.
DITTO Deep Colyn Engine, single.	11,18	7 10	2	28 3 0 28 0 0 10 0 0 35 0 0	18½ 18 19 10	Ditto	3696	292500	6 2	15357	22,135,167	6,77	Drawing perpendicularly 68 fathoms, and on the diagonal 9¼ fathoms. Main beam over the cylinder. One bob underground. BAWDEN.
DITTO Deep Danth, 30 inches, single.	14,31	8 0	3	29 0 0 15 3 0 40 0 0 16 0 0	10 11 7 8	Aug. 31st to Sep. 30th	1267	219670 156300	7 6	9768 6091	21,118,207	6	Drawing perpendicularly 44 fathoms, and on the diagonal 16 fathoms. Main beam over the cylinder, 32 fathoms of horizontal rods, and two bobs at the surface. BRATT.
DITTO South Mold, 60 inches, single.	16,1	6 6	3	56 1 0 16 0 6 11 4 0	12 10½ 8	Ditto	1705	307550 195590 239080	6 7	16342 3436 1527	23,339,316	7,46	Drawing in three shafts perpendicularly. Main beam over the cylinder. 280 fathoms of horizontal rods, and 4 bobs at the surface. BAWDEN.
HALKIN Party Gl. 50 inches, single.	9,31	6 0	3	84 3 0 11 0 0	10 8	Sep 1st to Oct. 1st	1622	333660	7 6 1 9	17383 1410	28,225,816	7,71	Drawing in three shafts perpendicularly. Main beam over the cylinder. 280 fathoms of horizontal rods, and 4 bobs at the surface. BAWDEN.
MHAVR 63 inches, single.	5,92	8 0	1	32 4 6 22 0 0	9 17	Ditto	1611	307970	8 0	18179	27,603,321	7,12	Drawing all the load perpendicularly. Main beam over the cylinder. 1/8 fathoms at horizontal rods, and bobs at the surface.
TALARGOCH 63 inches, single.	8,65	8 0	1	20 4 6 18 5 6	18 18½	Ditto	1501	185710	8 0	26892	26,716,505	4,27	Drawing all the load perpendicularly. Main beam over the cylinder. GODFREY.
GWERN-Y-MYNYDD 50 inches, single.	12,6	8 0	1	31 3 0 23 0 0 16 0 0 11 0 0	13½ 13½ 13 8 7½	Aug. 31st to Sep. 30th	2980	359120	7 0	28853	21,190,165	8,33	Drawing perpendicularly 52 fathoms, and diagonally 14 fathoms over a shorter. Main beam over the cylinder. WILLIAMS.

Printed by D. Batey, Rutheford-street, Holywell.

Stephen Eddy.

12. Eddy's Reporter, September 1829

70

The Engine Reporter probably began in 1827. It was compiled by Taylor's senior agent, Stephen Eddy, and printed by H.Davy of Whitford Street, Holywell. This single-sheet bulletin exactly followed the style of the Leans' publication, and had a similar title: 'Work performed by the following steam engines.' Duty was calculated on the basis of a bushel of coal. Almost certainly, this would have been taken as 84 lbs, then the legal equivalent weight[105].

By September 1829, ten single engines were at work, ranging from an 80-inch to a 36. The Captains (Stephen Eddy and Absolem Francis) were Cornish, as were most of the engineers (Bowden, Bratt, Godfrey and Grose) and the men. Many of them had gone by the time of the second known Reporter, February 1833, and only five engines were listed. Foreign competition had forced down the price of lead, but local landowners continued to demand excessive royalties, while petty squabbling between neighbouring adventurers hindered attempts at drainage.

No other Welsh Reporters have been found. If they were compiled, it was by another hand, because Stephen Eddy was sent away to manage Taylor's mines at Grassington in Yorkshire, and his career there illustrates the upward mobility that was possible for an intelligent miner. He had been born in 1800 into a humble mining family at St Just in Penwith, in the far west of Cornwall, and died a very wealthy, landed gentleman at Furness Abbey in Lancashire in 1861[93].

He was one of the official witnesses at the Wheal Towan trial in May 1828, and by September 1829 had been appointed Captain at Halkyn. He made his home at nearby Pant Glas, where his son, James Ray Eddy, was born in 1833. In the following year, he was sent to Grassington to replace the quarrelsome Captain John Barratt who Taylor transferred, first to his newly acquired copper mines at Coniston in Lancashire and then to Mexico.

Eddy spent the rest of his life in the north of England. At Grassington, the drainage problems were solved, not by steam, but by a series of leats which drove eight waterwheels. The only engine was a small portable, used for a short while in sinking a deep trial[94]. About 1840, Eddy supervised the building of a small multipurpose engine at the Cononley mine, but there are no records of its performance. These two instances appear to be the extent of his involvement with engines in Yorkshire[95].

As one of Taylor's senior and most experienced Captains, he was called upon to discuss drainage at the newly acquired Alport mines in Derbyshire, and, in the acrimonious disputes between the former management and the newcomers, he was described as 'an interfering little fellow'. Twelve years later, when Taylor withdrew from the area, Eddy became manager for the new company[96].

By now a man of some eminence, he had his fingers in a number of different pies. In 1852, he was appointed to assist John Higgins, the Crown Surveyor, in compiling a report on their Grinton mines in preparation for granting a new lease. Before this, he had helped to settle a dispute with the Earl de Grey regarding the Crown Mines at Askrigg Moor[97]. In 1854, he assisted Taylor with his properties in Cardiganshire, mapping the Cwmystwyth mines, and giving his name to Eddy's shaft at Logaulas[98]. Three years later, in 1857, he was invited to join the committee of management of a new company, formed to work the Rosewall and Ransom United Mines at Towednack in Cornwall. He probably agreed to the use of his name to help the initial flotation, but did not remain on the committee because of the difficulties of travelling between there and the North[99]. In that same year, he was appointed Agent to the prosperous Snailbeach Mining Company in West Derbyshire.

Stephen Eddy became Mineral Agent to the Duke of Devonshire, who in 1856/7 built a substantial house in the grand style for him on the estate. He named it Carleton Drive, and its land included Carleton Glen, a local beauty spot. The magnificence of the property showed the importance of his position within the ducal establishment. When he died in 1861 he left nearly £14,000 in cash, with a considerable number of expensive goods and chattels.

TAYLOR'S MEXICAN REPORTER

'Treasurer and administrator of the mines of the Union and others in Cornwall and Devon, of the "privileged mines" of the Duke of Devonshire and of the mines of the Earl of Grosvenor; as the inspector of the mines belonging to the Royal Hospital of Greenwich; and as a member of the council and the treasurer of the Geological Society of London.' This was how John Taylor identified himself to mining adventurers from Mexico[100].

The year was 1823: Taylor, aged 44, was at the peak of his career as the greatest mine manager of his day, and Mexico, after 15 years of civil war, had declared its independence from Spain. But the legendary silver mines had been devastated. The workings were drowned in water, shafts had caved in, timbers collapsed and buildings burnt. The new government tried to persuade its own citizens to invest in the mines, and when that failed, they revoked their traditional ban on foreign capital, and, almost overnight, Britain became a boom market for Mexican mining shares. John Taylor was foremost among those to seize the opportunity.

In 1824, he met the London representative of the Count of Regla, a wealthy Mexican landowner, and together they set up the Company of Gentlemen Adventurers in the Mines of Real del Monte. Among its objectives was the rehabilitation of certain mines 'by means of steam engines and other machinery'. Taylor's only knowledge of the country and its minerals was derived from Baron Humbolt, who had visited Real del Monte 26 years previously, in 1798, but his ignorance was matched by his enthusiasm. In uncharacteristic excitement, he took command of operations from his London office, but never found time to visit Mexico and see all the problems for himself.

He depended on receiving frequent progress reports, covering all aspects of the business. These were the responsibility of Captain John Rule, formerly of the United Mines, eldest son of the manager of Dolcoath, who Taylor had appointed his technical director in Mexico. He delegated the writing of the reports to Captain Hosking, but amended his subordinate's drafts just to prove that he was in control.

There were six steam engines at the mines: a 50-horsepower horizontal Taylor and Martineaux, three 30-inch, a 54-inch, and a 22-inch single that could be adapted to work as a double. In 1841, another 30-inch was installed. Because there was no coal, the engines were run on wood, cut from the surrounding forests, but supplies began to run short and had to be carted in from 20 miles away. The enormous tropical rainfall made the drainage of the mines of prime importance, and the shortage of fuel for the engines was a constant worry.

Taylor decided that, as well as the normal progress statements, he needed an Engine Reporter. In March 1831, he wrote

> I now receive a regular duty table from Mexico, showing the great advantage that the steam engines at Real del Monte have derived from the improvements to them[101].

These Reporters were probably prepared monthly, in similar format to those compiled in Wales for Taylor, and based on the style adopted by the Leans. The duty figures were not of course comparable with the British results since they were calculated, not on bushels of coal, but on cargas of timber. Like Lean's from 1833-6, they showed the amount of water extracted from each mine. There do not appear to be any copies in this country, but, according to Dr. Todd, who saw them in the archives at Real del Monte, they were published between 1833 and 1837, and I suspect for longer[102].

They may have begun in 1830. Taylor's comment that he 'now received' a duty report from Mexico was included in the extended version of his paper "On the duty of steam engines" completed on 22 March 1831. He did not refer to the Reporters in the original, shorter version of the text, published in 1829[103]. The implication is therefore that they began at some date between 1829 and March 1831, possibly nearer the latter since he 'now' received them. The post took three months to reach London, and they may have commenced at the end of 1830. There is no certain final date.

Taylor's enterprise in Mexico was a financial disaster, with a net loss of $4,451,158, setting investment of $15,381,633 against sales of silver amounting to $10,930,475. The balance of funds in London was down to £1,000 by February 1848 and no new source of ore had been discovered. No further calls could be made on shareholders because of the widespread panic induced by the revolutions in Europe. Internally, Mexico was in a highly turbulent condition, with rampant unrest and terrorism; externally in 1847, she was at war with America. Confidence, both in Taylor as a manager and in the prospects for mining on Mexico, plummeted. In October 1848, the company was wound up. Ironically, in less than 20 years, another British enterprise was returning handsome profits from these same mines, but Taylor's luck had run out and, even with his great managerial skills and experience, he had not succeeded in resolving problems that were greater and more complex than those that he overcame in the remote mining areas of England and Wales.

The Mexican fiasco did not discourage him from other foreign adventures. In the early 1850s he was engaged in Spain and California with the Nouveau Mond Gold Mining Company, and also in Spain with the Linares Lead Mining Company. No references to Engine Reporters have been found in connection with these projects, but, if engines were in use, it would be surprising that they were not as carefully monitored as at his other enterprises. Taylor was above all one of the earliest practitioners of scientific management.

THE DAY'S WORK

Amid all the talk about the engines it is easy to lose sight of the physical strain on people such as Lean, Browne, etc. in collecting the data and calculating duty. In many ways it required as much sustained effort from them as from the machines. The work had to be carried out week by week, month after month and, for the Leans, over many years. Whatever the weather or the state of their health, the Reporters were published.

The compiler of the statistics visited every mine on his list once in every four weeks. This involved long cross-country journeys, often on rocky, muddy tracks, in the prevailing Cornish wind and rain, sometimes in wheeled transport, more often on horseback. To some extent, the length of these monthly trips depended on the number of mines involved. This varied considerably; there were, for example, 59 listed by the Leans in 1832 and only 15 in 1852, but the mines were not necessarily close together, and having only a few to visit may have necessitated as much travel as twice or three times the number. These arrangements were constantly changing, and the only way to assess the workload is to take a snapshot at one moment in time.

In the late summer of 1829, Thomas Lean was recording the work of 43 engines at 28 mines. (Two years earlier, he and his brother John had jointly 52 engines at 38 mines, but they were now producing separate Reporters.) Thomas' mines were mostly centred around the Camborne-Redruth area, extending on the north side of the county to St Newlyn East, near Newquay; on the south, he had a long detour to visit two mines near St Austell. This was not an easy programme and it usually took him six working days to complete. The number of engines checked on any one day varied from three or four to as many as fifteen.

His travels can be tracked from the dates given in the Time columns of the Reporters. He did not always begin his trips on a Monday, and as a good Methodist, he did not work on Sunday. Sometimes he came home for the Sabbath; on other occasions he probably stayed with like-minded friends. During the course of the week away, he usually managed to spend one night with his family in the village of Praze-an-Beeble, south of Camborne, often after visiting Binner mine at nearby Leedstown. Taking one sample month his travelling arrangements were as follows.

Itinerary, 26 August - 2 September 1829

	Place	*Mine*
Wednesday 26 August	Praze-an-Beeble	—
	St Ives	St Ives
	Lelant	Lelant
	Perranzabuloe	Prosper
		Rose
	St Newlyn East	Deer Park
Thursday 27 August	Breague	Great Work
	Perranuthnoe	Caroline (2)
	Gwinnear	Hope
	Leedstown	Binner Downs (2)
Friday 28 August	Camborne	Dolcoath
	Scorrier	Montague
	Redruth	Harmony
		Cardew
Saturday 29 August	Redruth	Tolgus
	St Blazey	East Crinnis (2)
	Par	Pembroke (2)
Sunday 30 August	—	—
Monday 31 August	St Day	Unity (2)
		Poldice (2)
		Damsel (2)
		Ting Tang (2)
	Wendron	Beauchamp
	Gwennap	Tresavean
Tuesday 1 September	Gwennap	Consols (6)
		United (2)
	Kenwyn	Falmouth
		Sperris
Wednesday 2 September	St Day	Penrose
	Praze-an-Beeble	—

N.B. Numbers in brackets are the numbers of engines checked at each site.

Week 1 began on Wednesday, 26 August 1829. There can be no certainty about the order in which he visited the mines on that, or any, day, but it is certain that he visited that group. Plotting the sites on a map suggests how he might have organised his travelling, and suggests where he may have stayed overnight. The first day, he went down to St Ives and then north to St Newlyn East. In the morning, back west, and by the end of Thursday he was not far from Praze, enabling him to sleep at home. On Friday, 28th, he went to the Camborne-Redruth area, probably spending the night there, since he went to another nearby mine on the Saturday before departing eastwards towards the St Austell area, where he spent Sunday. Monday was a heavy day, inspecting eight engines at St Day before going down to Wendron and back to Gwennap, where on the Tuesday he checked ten engines, before going back to St Day before heading home.

He had three clear weeks to do his calculations, get the text to the printer at Camborne, correct the proofs, and send the copies off to his subscribers. Then, on Friday, 25 September, his journeys began again, spending the night at Perranzabuloe. Back home for Sunday; Camborne-Redruth on Monday 28th; Tuesday, spent travelling to St Bazey and Par; Thursday at St Day, Kenwyn and Gwennap (fifteen engines); Friday, after finishing at Gwennap, back to St Day, to Wendron, and home. Three weeks later, on 27 October, he set off for St Ives at the beginning of yet another circular trip. Comparison of several Reporters around this time shows that his route did not greatly vary, except for including Penrose with the others at St Day.

These were very demanding schedules, requiring great physical resilience and enthusiasm to carry them out with unfailing regularity every four weeks throughout that and every succeeding year. Thomas Lean senior did it from 1811 to 1847, by which time he was 63 years old. His son performed the work from 1847 until, probably, 1897, when he was 70. John and Joel Lean helped for much shorter periods. William Tonkin reported for seven years, and William Browne for twelve. It was the two Thomas Leans, father and son, who bore the brunt of the work.

For each pumping engine, he confirmed the load p.s.i. on the piston, and noted the number of strokes recorded by the counter; he examined the coal-measurer's lists so as to calculate the amount of fuel used. He was then able to work out the duty. (The methodology of this is explained in Appendix 1.) Where there were stamps, he noted the coal consumption, the number of strokes and the number per minute. For whim engines, he needed to know the amount of coal consumed, the total number of kibbles drawn, and the number drawn per measure of coals.

All these figures were carefully entered onto a master copy of the previous month's Reporter, with the old figures neatly crossed out and replaced by the new data. Complete new entries were made for new subscribers, and the records of non-participants deleted. The date of each visit was added in the Time column. For example, if the previous month's Reporter had given 'Aug 29th to Sep 28th', the August date was deleted and the new one added at the bottom, this becoming 'Sept 28th to Oct 29th'. It was sometimes necessary to amend the Remarks column, and often needful to add or remove notes regarding the performance of a particular engine during the month under review: for example 'Polladras Downs engine, and Wheal Tolgus engine, have been working with leaky boilers' (Lean, January 1833) or 'the shaft-work of Wheal Leisure engines have been altered this month' (January 1835). When all this had been done, the pages were delivered to the printer, who seems likely to have kept the type in a standing forme. The Reporters all used convenient local commercial printers, except the Leans who after 1853 maintained their own press until Thomas the younger retired, when it was sold.

Browne allowed himself a slightly less punishing travel routine than Lean. For one thing, his mine visits were not always at exact four-weekly intervals, but the dates were given clearly on his Reporters. He was lucky also that his home base at St. Austell was roughly midway between the furthest west that he had to go, near Camborne, and his eastern mines, over the Devon border. This allowed him to plan his schedules in two parts, but there was probably little real difference in the overall length of his and Lean's journeys.

One significant difference between the two Reporters was that almost every issue of Browne's had an explanation of some part of the methodology used in the calculations. He and West sought consensus from the mining community, and it would have been the engineer who laid down the standards to be used, leaving Browne to deal with the correspondence and the day-to-day routines.

Can we look for any common characteristics in the men who served as registrars and reporters of engines? Did they have similar backgrounds or shared experiences? On the whole, they did not. Most of them were Nonconformist in religion, but this is not really relevant. Some were engineers: Joel Lean the elder and his second son Joel, William Tonkin and James Keast, but neither of the two Thomas Leans nor John Lean appear to have had any real technical knowledge. Some had actually worked in the mines, like old Joel Lean, Stephen Eddy in Wales and the compiler of the Mexican reporters, but many of the others had not been underground. The only thing that seems to unite them is a facility with figures and a methodical mind. Integrity? Not Thomas Lean the elder, nor Tonkin, and there is some doubt about William Browne. A passionate desire to improve engineering practice? Only old Joel Lean.

They were all very different people, with very different motives for doing the job, but we owe a debt of gratitude to those men who travelled the county in all weathers, month by month and year by year to record the performance of the Cornish engines.

APPENDIX 1: THE MEASUREMENT OF DUTY

Mechanical power can be described in various ways. The usual 19th century custom was to use horsepower as the measure of the effect of rotary engines as used, for example, in the factories of the Midlands and the North, whereas the power of Cornish pumping engines was expressed as their Duty. Horsepower measured performance against time: the amount of work done in a given period. Duty measured performance by the amount of fuel used: the amount of work performed by a steam engine for the combustion of a given amount of coal.

The concept of 'duty' as a unit of efficiency evolved during the latter part of the 18th century. John Smeaton measured engine performance by what he called 'Effect': the volume of water lifted one foot each minute for burning a bushel of coal each hour in the boiler. It had the dimensions of 'cylindrical inch feet squared per minute per bushel per hour' and used an 88 lbs bushel of coal. About 1779, James Watt (1736-1819) was describing efficiency as cubic feet of water lifted a foot for burning a bushel of coal, or millions of pounds of water lifted a foot for burning a hundredweight of coal (at 120 lbs). He seems to have adopted the word 'duty' by the end of the 18th century, and by about the year 1800 duty became the accepted dynamic unit[104].

The people at the time knew just what quantities were being used, or else defined them in correspondence. They did not feel it necessary to record the details for posterity, and the matter can get very complicated. It must always be remembered that an English bushel may be 79, 84, 88 or 94 lbs; a hundredweight might be 100, 112 or 120 lbs.

Duty was the number of pounds (expressed in millions) lifted one foot high by the combustion of a certain amount of coal, and that amount differed in different Engine Reporters and at different times. The Leans used the 84 lbs bushel from 1811 - 1835 and the 94 lbs bushel until 1847; from then until 1856, they expressed duty both in terms of that bushel and of the 112 lbs hundredweight; from July 1856 they only used the 112 lbs hundredweight[105]. William Tonkin's Engine Reporter used the 84 lbs bushel from 1834 to August 1835, when he transferred to the 112 lbs hundredweight. John Taylor's Welsh Reporters used the 84 lbs bushel. William Browne used the 112 lbs hundredweight throughout (1847-58).

Neither the Leans nor any of the other engine reporters explained

Pumping Engines—Continued.

MINES.	Time.	ENGINES.	Length of the stroke in the cylinder. (Feet)	Length of the stroke in the pump. (Feet)	No. of lifts.	Depth. (Fms. ft. inches)	Diameter of the pump. (inches)	Load in pump. (Lbs.)	Load per square inch in the piston. (Lbs.)	Number of Strokes.	Number of strokes per minute.	Consumption of coal in bushels. (Bush.)	Pounds lifted one foot high, by consuming a bushel of coal.	Average quantity of water drawn per minute. (Imp. Gals.)	REMARKS, and ENGINEERS' NAMES.
UNITED MINES.		Taylor's, 85 inches single.	11,0	10,0.	5 1 1	156 2 / 12 0 / 4 3	14 13 7	67270	10,78	191200	4,6	1303	98,859.570		Drawing perpendicularly. Main beam over the cylinder, two balance bobs at the surface, and two ditto underground. Hocking and Loam.
	June 2 to July 1.	Carloon's, 90 inches single.	9,0	8,0.	5 2	156 0 / 28 5	16 14	93214	13,0	220000	5,4	3053	53,219,258		Drawing perpendicularly. Main beam over the cylinder, and one ditto under ground. Hocking and Loam.
		Eldon's, 30 inches single.	9,0	7,5.	1	34 0	14	13631	16,0	431200	10,3	715	61,023,553	1487,5	Drawing perpendicularly. Main beam over the cylinder. Hocking and Loam.
		Lean's, 85 inches single.	10,0	7,5.	1 4 1	7 1 / 139 4 / 12 5	12 17 16½	91602	12,2	293300	7,07	3026	67,044,127		Drawing perpendicularly. Main beam over the cylinder, one balance bob at the surface, and one ditto underground. Hocking and Loam.
		Hocking's, 85 inches single.	10,0	8,5.	6 1 1 1	170 3 / 14 2 / 13 5 / 8 5	16 16½ 16½ 7	103692	15,33	270800	6,6	3529	69,131,761		Drawing perpendicularly. Main beam over the cylinder, one balance bob at the surface, and two ditto underground, and an Air Machine 10 and half inches diameter. Hocking and Loam.
ROSSOE BRIDGE.	ditto	50 inches single.	10,0	8,0.	1 1 1 1	23 0 / 10 0 / 10 0 / 10 3	14 12 9 14	18100	7,33	153500	3,65	409	54,477,451	195,1	Drawing the load in two shafts perpendicularly. Main beam over the cylinder. Two bobs and 70 fms. horizontal rods at the surface, and 40 fms. dry rods in the shaft. P. Michell.
SOUTH WHL. TOWAN.	June 9 to July 9.	70 inches single.	10,0	7,5.	2 1 1	71 4 / 11 4 / 11 3	16 15 12	46195	9,0	219010	5,00	2042		331,17	Drawing perpendicularly 60 fathoms, and the remainder diagonally. Main beam over the cylinder, balance bob at the surface, and one angle sheave underground. J. West.
UNITED HILLS.	ditto	Williams's, 80 inches sngle.	10,0	8,0.	1 1 1 1	41 4 / 30 2 / 25 0 / 10 5	13 16 16½ 12	50922	8,1	188840	4,36	1208	59,267,244	410,38	Drawing 12 fathoms perpendicularly, and the remainder diagonally. Main beam over the cylinder, and one angle bob in the shaft. J. Sims.

13. Extract from Lean's Reporter, June 1841, showing statistics for Taylor's engine at United Mines

how they were calculating duty, and none of them seems to have had the word in the main body of their text. The Leans did not specify the weight of the bushel until 1838, when they gave it as 94 lbs, but when they changed to the hundredweight, always cited it as being 112 lbs. In the different series of Reporters (the Leans', Browne's, Taylor's) the duty figure was invariably given as: millions of lbs lifted one foot high by consuming [quantity] coal.

The first time that the Leans printed the word duty was in the Engine Reporter for October 1838 as part of an ancillary table summarising the month's results. They employed the word again in July 1856 as part of an explanatory note when they changed the coal measure. With a few exceptions, it was not a term that they liked.

The method of calculating duty was complex. First, the diameter, in square inches, of the plunger piston was multiplied by the height of the lifts in fathoms and by 2.045. (This last was taken as the weight in pounds of a cylinder of water one inch in diameter and six feet long[106]).

The result was then multiplied by the length of stoke in the pumps and by the number of strokes made in a given time. This gave the number of pounds lifted one foot high during the period under review (usually four weeks).

This figure was then divided by the number of bushels (later, hundredweights) of coal consumed. The quotient was the duty.

Taking as an example, Taylor's engine with seven pumps at United Mines, as reported in June 1841. The calculation was

a) 5 pumps each 14 inches in diameter, together 156 fathoms 2 ft in length
   $\therefore 5 \times 156^2/_6 \times 14^2 \times 2.045$ = 62,674 lbs

b) 1 pump, 13 inches diameter, 12 fathoms lift

   $\therefore 1 \times 12 \times 13^2 \times 2.045$ = 4,146 lbs

c) 1 pump, 7 inches diameter, $4^1/_2$ fathoms lift
   $\therefore 1 \times 4^1/_2 \times 7^2 \times 2.045$ = 450 lbs

   Total load = 67,270 lbs

d) the length of the stroke in the pumps was 10 ft; the number of strokes from 2

   June to 1 July was 191,200 and the coal used was 1,305 bushels
   $\therefore \dfrac{67,270 \times 10 \times 191,200}{1,305}$ = 98,559,570 lbs

   Thus giving a duty of 98.5 million[107].

The difficulty with the calculation was to count the number of strokes made by the engine. This was not a problem with the very earliest engines that would have made only about 12-15 per minute, but when higher-speed machines were developed, a special instrument was required. This was known as a Counter. It recorded every movement from the horizontal made by the great lever. In essence, the device was a series of wheels and pinions, set in motion by a pendulum, acting through an escapement similar to that in a longcase clock; the number of movements made by the pendulum were registered by a train of interconnected dials. These workings were enclosed within a box that was fastened to the main beam of the engine.

The counter was invented by John Whitehurst (1713-86), a Derby clockmaker, and improved by Watt, but when he found that the dials could be reset by persons wishing to avoid paying the premium, he introduced a second model in a tamperproof box. These were manufactured at the Soho works. A third version was designed by Arthur Woolf in 1810/11, and was secured by patent Bramah locks, made by Joseph Bramah, for whom he had worked in London. A later style, known as Harding's Improved Counter, was made by Harding, Leeds and James Simpson of London. They made the box from brass: Watt's had been wooden, and Woolf's iron.

Similar precision could not be achieved in measuring the coal, and the quantities were not necessarily always measured exactly in the hurly-burly of daily life at a mine. There were other circumstances that made it impossible to judge the exact weight of coal. A great deal of rain falls in Cornwall: wet coal, particularly if it is in small pieces, weighs more than coal in large lumps. Sometimes furze or wood might be burnt to supplement the coal, or the ashes might be raked through to find cinders that could be reused. Tricks could be played with the engine. It could be run on a short stroke, when no one was about, say, at night or in the weekend. If pitwork were neglected, more air than water might be drawn up with correspondingly improved performance, on paper, but not in effect.

Duty was subject to a number of variables. At best, it was achieved honestly by men of integrity. At worst, it was falsified as a result of dubious practices or the downright intention to deceive. It is not always obvious from the columns of the Reporters whether the statistics are an accurate reflection of an engine's performance. Generally, the figures can be relied upon, but if there was a great deal at stake, by way of money or personal reputation, the temptation to exaggerate the duty may sometimes have been too much to resist.

Suspicions lingered throughout the duty wars, but by the mid-19th century attitudes had changed. Economy became more important than prestige. As early as 1859, Wicksteed suggested the duty per pound sterling was as important as the duty per hundredweight of coal[108]. Lean did not adopt the idea, but the days of glory were coming to an end.

APPENDIX 2: COPIES OF THE REPORTERS

1. Lean's Reporter
Cornish Studies Library, Redruth

1812	May
1813 - 1880	probably complete
1881	February, June, August, October - December
1882	February, April - June
1883	March
1884	March, June, October
1885	January, March, June, August, October
1886	April
1888	December
1889	January - December
1890	January - March, May, August, November, December
1891	January, July, August, October, December
1892	January, March, May-August
1893	lacks April
1894	January - December
1897	April - July, September
1898	March, April, July - December
1899	January - November
1900	May, December
1901	January, April
1904	May, July

Science Museum Library (on microfilm and hard copy)

1812	January
1814	March
1815	July
1817	April, May, July
1818	January - December
1819	March - August
1820	January - December
1826	January - December
1829 - 1832	lacks December 1830
1834 -1838	lacks November 1835; January and July 1836; August 1838
1839 - 1846	lacks September 1840; January 1843

1848 - 1851	lacks January - May 1849
1853 - 1855	lacks June 1854
1863 - 1864	January, March 1863; November, December 1864
1865-1869	lacks December 1865; August, November 1866; January 1869
1871	lacks January, February
1872 - 1873	lacks January, April, June, November 1872; December 1873
1875 - 1877	lacks October 1875; March, October 1876
1895	January, March, April
1896	November, December
1901	April

Science Museum, Goodrich collection (item 784)

1818	May, June
1825	June, November
1827	August
1829	October

British Library

| 1838 - 1855 | June 1838 - December 1855 |

2. John Lean's Reporter
Cornish Studies Library (on microfilm)

1827	September - December
1828	January - December
1829	lacks April
1830	January, February, March, August
1831	February. March proof copy

Science Museum Library (on microfilm and hard copy)

| 1828 | lacks March. September is in the Goodrich collection |
| 1830 | lacks July and September. March and April in Goodrich collection |

3. Tonkin's Account
No copies have been found

4. Browne's Reporter

The only surviving issues appear to be those for January - September 1847 (no.1 - no.8), January - August 1848 (no.12 - no.19), and December 1854 (no.95) These are on microfilm at the British Library. All except the last (1854) are also on microfilm at the Cornish Studies Library. Monthly summaries can be found in the *Mining Journal,* 1850 (2 February, p53) to 1858 (12 June, p391) but with some omissions.

5. Taylor's Welsh Reporters

Copies of the only two known issues have been deposited in the Welsh National Industrial and Maritime Museum, Cardiff. There is a copy of September 1829 in the Goodrich collection (item 784a), Science Museum; February 1833 is in the Howarden Office, Clwyd Record Office.

6. Taylor's Mexican Reporters

No copies have been found in the U.K.

APPENDIX 3: REFERENCES

1. Reprinted in GALLOWAY, R.L. *The steam engine and its inventors.* (1881, Macmillan) p.99
2. HOWARD, B. "The duty on coal, 1698-1831". (*Journal of the Trevithick Society*, no. 26, 1999, pp.30-5)
3. Reprinted in FAREY, J. *Treatise on the steam engine: historical, practical and descriptive.* (Vol. I, 1827; vol. II not originally published. Both reprinted 1971, David & Charles), vol. I, p.183
4. FAREY, J. (op.cit.), vol. II, p.92, footnote
5. GILBERT, D. "On the progressive improvements made in the efficiency of steam engines in Cornwall". *(Royal Society, Philosophical Transactions,* 1830, pp.121-32)
6. TODD, A.C. *Beyond the blaze: a biography of Davies Gilbert.* (1967, Barton) pp.63-4
7. GILBERT, D. (op.cit.5), p.126
8. LEAN, T. and brother. *Historical statement of the improvements made in the duty performed by the steam engines in Cornwall from the commencement of the publication of the monthly reports.* (1839) (Reprinted 1969, Barton, as *On the steam engines in Cornwall*) p.7
9. HARRIS, T.R. *Arthur Woolf, the Cornish engineer, 1766-1837.* (1966, Barton) pp.50-1
10. LEAN, T. (op.cit.8), p.8
11. BOAS, G.C. and COURTNEY, W.P. *Bibliotheca cornubiensis.* (3 vols., 1874-82) p.308
12. LEAN, T. (op.cit.8),p.10
13. [LEAN, T. ?] "Some observations on steam engines". (*Philosophical Magazine*, 1815, vol.46, July, pp.116-20, 462-4) pp.116-9
14. FAREY, J. (op.cit.3) vol.II, p.93
15. (op.cit.8)
16. (op.cit.8) pp.10-1
17. *Philosophical Magazine*, 1802, vol.17, p.40; 1803, vol.19, p.133; 1804, vol.22, p.123; 1815, vol.46, pp.120-2 and pp.43-4
18. [LEAN, T ?] (op.cit.13) p.119
19. FAREY, J. (op.cit.3) vol.II, p.91
20. WICKSTEED, T. "On the effective power of the high-pressure expansion condensing steam engines commonly in use in the Cornish mines." (*Institution of Civil Engineers, Transactions,* 1836,

vol.I, pp.117-130) p.120
21. VALE, E. *The Harveys of Hayle.* (1966, Barton) p.90
22. (op.cit.8) pp.10-1
23. For example, *Philosophical Magazine,* (op.cit.13) p.119
24. BARTON, D.B. *The Cornish beam engine* (1965, 2nd ed. 1969, Barton) p.34, fn.2. HENWOOD, W.J. "Account of the steam engines in Cornwall". (*Edinburgh Journal of Science,* 1829, vol.X, pp.34-49) p.36
25. Science Museum Library, London. Goodrich collection, item 784/7
26. 1817, House of Commons paper 442
27. *Philosophical Magazine,* 1815, vol.46, August, pp.116-20
28. *Philosophical Magazine,* 1815, vol.46, December, pp.462-6
29. HENWOOD, W.J. (op.cit.24)
30. *Philosophical Magazine,* (op.cit.28) pp.460-1
31. BURTON, A. *Richard Trevithick: giant of steam* (2000, Aurum press) p.170
32. *Philosophical Magazine,* 1818, vol.49, p.465
33. FAREY, J. (op.cit.3) vol.II, p.101, 122-5
34. FAREY, J. (op.cit.3) vol.II, p.125, fn (a)
35. FAREY, J. (op.cit.3) vol.II, p.296
36. FAREY, J. (op.cit.3) vol.II, p.215
37. TAYLOR, J. "On the duty of steam engines". (*English's Quarterly Mining Review,* 1831, pp.33-57) p.56
38. HARRIS, T.R. *Arthur Woolf, the Cornish engineer 1766-1837.* (1966, Barton); HOCKING, S. "A brief sketch of the life and labours of Arthur Woolf, engineer". (*Miners' Association of Cornwall and Devon, Report and Proceedings,* 1875, pp.8-22)
39. FAREY, J. (op.cit.3) vol.II, p.191
40. TAYLOR, J. (op.cit.37) p.54
41. *Royal Cornwall Polytechnic Society Report:* 1835, pp.136-7; 1836, pp.142-3; 1837, pp.78-9; 1838, pp.169-70. *Mining Journal,* 1839, p.22; 1840, p.39
42. *R.C.P.S. Report,* 1836, p.34
43. FAREY, J. (op.cit.3) vol.II, p.232
44. Weights and Measures Act 1834 (4&5 Will.IV, cap.49)
45. POLE, W. *A Treatise on Cornish pumping engines.* (1844-8, Weale) p.157
46. LEAN, T. (op.cit.8) pp.143-5

47. FAREY, J. (op.cit.3) vol.II, p.252
48. HENWOOD, W.J. "Statistical notices of the mines in Cornwall and Devon". (*Royal Geological Society of Cornwall, Transactions* 1843, vol.5, pp.461-82)
49. TAYLOR, J. "On the duty of steam engines." (*Records of Mining*, 1829); *Philosophical Magazine*, 1830, pp.424-31; *Philosophical Magazine*, 1836, pp.67; *English's Quarterly Mining Review*, 1831, pp.33-57; *British Association, Report and Transactions*, 1835, pp.108-9
50. British Association archive, held in the Bodleian Library, Oxford. Some documents are missing and the exact date of the application for a grant cannot be determined, but was probably late in 1837 or early in 1838.
51. LEAN, T. (op.cit.8)
52. TAYLOR, J. (op.cit.40)
53. A.D.1812, No. 3621
54. A.D.1825, No. 5228
55. HUMPHREYS, E. *Reminiscences of Briton Ferry and Baglan* (1898, Swansea) p.18-9; INCE, L. *The South Wales iron industry* (1993, Ferric Publications) pp.95-6
56. *Athenaeum: journal of English and foreign literature, science and the fine arts,* no.627, 2 November 1839, pp.822-3
57. POLE, W. (op.cit.45) p.149
58. Copies addressed to them are in the Goodrich Collection, Science Museum, London
59. TODD, A.C. (op.cit.6) p.99
60. POLE, W. (op.cit.45) p.149, fn.41
61. *Mining Journal,* 1857, 14 October, p.722
62. For example, TRESTRAIL, N. "The duty of Cornish pumping engines". (*Federated Institute of Mining Engineers, Transactions* December 1896, pp.548-62) p.552
63   TEAGUE, W. "On pitwork and the duty of Cornish engines". *(Proceedings of the Mining Institute of Cornwall,* 1882, vol.1, no.6, pp.188-91)
64. TRESTRAIL, N. (op.cit.62) p.558-9
65. MARAZION HISTORY GROUP. *The Charter town of Marazion.* (1995); *Cornish Post and Mining News,* 16 August 1908; *The Cornishman,* 16 August 1908; *West Briton,* 16 August 1908;

memorial panel in Marazion Council Chambers
66. BARTON, D.B. (op.cit.24) p.79
67. Information from Mrs. Margaret Bunney
68. *Institution of Mechanical Engineers, Proceedings*, 1914, p.1012. Obituaries in the *Cornish Post and Mining News*, 4 June 1914, p.5, and in the *West Briton*, 4 June 1914, p.4
69. Information from Mrs. Margaret Bunney
70. FAREY, J. (op.cit.3) p.235
71. BOAS, G.C. and COURTNEY, W.P. (op.cit.11) p.1346
72. HENWOOD, W.J. (op.cit.48) p.462
73. FAREY, J. (op.cit.3) vol.II, p.235
74. op.cit. 43
75. FAREY, J. (op.cit.3) p.252
76 FAREY, J. (op.cit.3) p.235, 238, 252; LEAN, T. (op.cit.15) pp.143-5
77. HALLETT, M. *Portsmouth water supply, 1800-1860* (Portsmouth paper no.12) (1971 Portsmouth Museum) pp.19-20
78. Portsmouth Water Company archives. (Portsmouth Museums and Records Service)
79. Cited in MICHELL, F. *Annals of an ancient Cornish town.* (1978) p.119
West Briton, 6 November 1835, and successive issues. Cited by Barton (op.cit.24) p.49, fn.4
80. BARTON, D.B. *A History of copper mining in Cornwall and Devon.* (3rd ed. 1978, Barton) p.58
81. BARTON, D.B. (op.cit.24) p.32, fn.1
82. Letter, Davies Gilbert to Thomas Telford, 4 June 1825 (Royal Institution of Cornwall, Trevithick papers)
84. MORSHEAD, W. "On the duty of the Cornish pumping engines". *(Institution of Civil Engineers, Minutes of Proceedings, vol.23, pp.45-85)* p.46. Barton (op.cit.24, p.54, fn.2) probably incorrectly attributes this to William Husband, without providing a citation.
85. A.D.1849, No. 12,789
86. *Royal Cornwall Gazette,* 9 May 1851, p.5 and 2 May 1862, p.4
87. Cornwall Record Office, CF 3907 and CF 3904
88. *Mining Journal,* 1857, 29 August, p.611
89. Letter quoted by Barton (op.cit.24) p.54, fn.3
90. TRESTRAIL, N. (op.cit.62) p.552
91. TAYLOR, J. (op.cit.37) p.57

92. See BURT, R. *John Taylor, mining entrepreneur and engineer, 1779-1863.* (1977, Morland) pp.22-4 and BAYLES, R. "A brief survey of beam pumping engines employed on lead mines in Flintshire". (*Northern Cavern and Mine Research Society, Memoirs*, 1968, January, pp.8-14)

93. GILL, M.C. *The Yorkshire and Lancashire lead mines.* (British Mining, no. 33) (Northern Mine Research Society, 1987) pp.26-7

94. BURT, R. (op.cit.92) p.27

95. Private communication, Mike Gill

96. BURT, R. (op.cit.92) p.55

97. Private communication, Mike Gill. See also, Public Record Office, CRES/61, cited in TYSON, L.O. and SPENSLEY, I.M. *The Grinton mines.* (British Mining, no. 5) (Northern Mine Research Society) pp.46-7

98. Private communication, R.P.Jones, National Museum of Wales

99. Private communication, Justin Brooke

100. RANDALL, R.W. *Real del Monte: a British mining venture in Mexico.* (1972, Texas university press) p.36. In this chapter, I have drawn on this book and on TODD, A.C. *The Search for silver.* (1977, Lodeneck press, Padstow)

101. TAYLOR, J. (op.cit.37) p.57

102. TODD, A.C. (op.cit.100) p.123

103. TAYLOR, J. (op.cit.49)

104. Private communication, Dr. J. H. Andrew

105. HOWARD, B. "Was the bushel 84 lbs or 94?". *(Journal of the Trevithick Society*, no. 29, 2002. pp123-8)

106. This was spring water which weighed less than that extracted from the mines, but the excess was notionally set against any other possible errors

107. POLE, W. (op.cit.45) p.148

108. WICKSTEED, T. *Further elucidations on the useful effects of Cornish pumping engines.* (1859, Weale) p.8

Acknowledgements

Among the many people that have helped me, I would like particularly to thank Dr. J. H. Andrew, Kenneth Brown, Margaret Bunney, Chris Hodrien, Rodney Law, and Don Norman. Also, Kim Cooper of the Cornish Studies Library, the Athenaeum Club, the Bodleian Library, Clwyd Record Office, Cornwall Record Office, House of Commons Information Office, Portsmouth Museums and Records Service, Robert Sharp at the Science Museum Library, and many others. Finally, I am grateful to the authors of the books that I have cited.

Index

Armstrong, William 47
Austen, J. T. *see* Treffry, J. T.

Barratt, John 71
Beighton, Henry 9
Boiler explosions 24-5
Boulton, Matthew 11
Boulton & Watt
  engines 9-12, 20
  profit 12
Bramah, Joseph 19, 21, 31
British Association 40, 42
Briton Ferry 44
Browne, William
  career 59, 64, 67
  character 7, 80
  data collection 76, 79
  family 59
Browne's Reporter 7, 50, 59-68,79, 81
  advertisements 64
  allowances 50, 63-4, 67-8
  and R.C.P.S. 50
  cost 62
  criticism of 61, 64, 67-8
  data collection 78, 79
  summaries 62, 66
  West's involvement 55, 64-68
Budge, John 59
Bull, Edward 15
Bushels
  changed 83
  Browne 61, 62
  Lean 25, 37, 44, 48
  Tonkin 56
  Welsh 71

Camborne School of Mines 53-5
Civil Engineers, Institution of 39, 47
Coal
  measurement 39, 63, 84
  quality 64
  saving 9, 11, 20, 27, 85
  tax 9
Commons, House of 24-5
Cornish Miners' Society 46
Coulson, Thomas 24, 36
Counters 84
  accuracy 46-7
  Browne's 62
  locks 21, 31, 84
  tampering 31, 84
  Watt's 11, 84
  Woolf's 21, 84

Davey brothers
  Lean, Joel 18,21-3
  Reporters 18-9, 21, 22-3
Davey, John 11, 13. 15, 18. 27
Davey, William 18, 20, 21, 27
Duty
  calculation of 44, 78, 81-5
  definition 7, 83
  falsification 27, 28-30, 31, 33,
    37, 40, 42, 84
  highest 7
  vs. horsepower 37, 47, 62, 64
  plegometer 46
Eddy, Stephen 69-72, 80
Eddy's Reporter 69-72
Edwards, Humphrey 19
Engine trials *see* Trials

Engineers
    named in Reporters 24, 51
Engines
    neglect 12, 13, 51, 53
    number of 39, 56

Farey, John
    at Abraham 28-9
    Lean, John 36
    Tonkin 56
    Woolf 20, 33
Francis, William 29

Gilbert , Davies 11, 47
Goodrich, Simon 24,36
Great Exhibition 67
Grose, Samuel
    Browne 59
    Towan 29, 30, 59
    Woolf 33

Harris, Blanche 13, 14
Henwood, William 27, 39, 56
Historical Statement
    advertised 45-6
    authorship 18, 40-2
    commissioned 40
    timing 19
    truth 39, 21-3, 42
Hocking & Loam 59, 61
Hornblower, Jonathan 11, 15, 42
Horsepower
    in Reporters 62, 64
    vs. duty 47,81
Horses
    vs. engines 37
Hosking, Captain 73

House of Commons 24-5
Hundredweight 61, 62, 81

Keast, James Champion 53-5, 79

Lean, Joel (elder) 13-7
    career 13-17, 79-80
    character 13, 79-80
    Davey brothers 18, 21-2
    family 13
    Reporter 15-17, 22
Lean, Joel (younger)
    career 42-3, 48, 79
    family 16
    Historical Statement 40, 42-3
    Reporter 44-8, 78
    Taylor 42-3
    technical knowledge 42, 79
Lean, John
    family 16
    Reporters 7, 18, 30, 31-36, 56-
        7,68,78
    Quaker 31, 34
    technical knowledge 79
    Thomas Lean 7,22,24,30,55
Lean, Thomas (elder)
    and Joel Lean (elder) 22
    and Joel Lean (younger) 18,
        40-3
    and John Lean 18, 22, 24, 31,
        55, 68
    career 47, 78
    data collection 76-9
    dishonesty 7, 22, 27-9, 30, 31-2,
        33, 36, 43, 61, 80
    family 16, 47
    Historical Statement 22-3, 39,
        40-43

Reporter 7, 24-30, 34, 37-43, 68,
   77
supports Woolf 18, 22, 24-7, 28-
   30, 31, 36, 37
technical knowledge 24, 79
Lean, Thomas (younger)
   career 48, 52-3
   character 7
   family 16
   Reporter 48-53
   visited America 51-2
   technical knowledge 79
Lean's Reporter
   advertisements 45-6
   allowances 50, 68
   beginning 12, 14-17, 18
   cost 62
   coverage 51, 52, 68
   data collection 76-9
   decline 53, 55
   editorial work 78-9
   format 15, 24, 37, 44-5, 48, 50
   influence 7, 45, 47
   masthead 44, 50
   numbering 46
   printing 48, 55, 79
   purpose 7, 17, 20, 47
   rainfall table 48-50
   summaries 25-6, 42, 46, 62, 66
   Taylor 37-43
   title changes 44-5
   water extraction table 37
   water meter table 37-9
   see also   Lean, Joel
           Lean, John
           Lean, Thomas
           Taylor, John

Woolf, Arthur
Loam, Matthew 12, 33, 67
Marazion 47, 48, 52-3
Mines
   Abraham 21, 24, 25, 27, 28
   Alfred 15, 29
   Alport 71
   Anna 67
   Askrigg Moor 72
   Balnoon 30, 36, 56
   Bassett 53
   Beauchamp 77
   Binner Downs 76-7
   Cardew 77
   Carn Brae 53
   Caroline 77
   Chance 9, 28
   Consols 22, 29, 31, 40, 77
   Cook's Kitchen 28
   Crenver 13, 15, 27
   Crinnis 30, 34, 36
   Cwmystwyth 72
   Damsel 77
   Deer Park 76
   Devon Gt. Consols 67
   Dolcoath 9, 28, 30, 61, 73, 77
   Drakewalls 61
   East Crinnis 77
   Falmouth 77
   Fortune 24
   Fowey Consols 7, 46, 59, 61
   Godolphin 56
   Great St. George 30
   Great Work 77
   Grinton 72
   Gwernymyndd 69

Halkyn 69, 71
Harmony 77
Herland 11
Hope 77
Lelant 77
Llynpandy 69
Logaulas 72
Maid 9
Milwr 69
Montague 77
Oatfield 13, 15,27
Pembroke 77
Penrose 77
Penwith 56
Perran 30, 56
Poldice 9, 77
Polgooth 59, 67
Polladras 30, 34, 36, 56, 79
Prosper 77
Real del Monte 73-5
Reeth 30, 34, 36
Rose 30, 77
Rosewall 72
Snailbeach 72
Sperris 77
St. Ives 30, 77
Stray Park 24, 30, 34, 36, 56, 68
Tincroft 52
Ting Tang 77
Tolgus 77
Towan 29, 30, 36, 40, 56, 71
Tresavean 56, 77
Treskirby 28
United 28, 29, 46, 73, 77
United Hills 30, 56
Unity 14, 77
Virgin 9

Vor 25, 27, 28, 30, 34, 36
Wellington 30
Mexico 73-5
Mining depression 50-1
Monthly Duty Paper 40
Morshead, William 65

Neath Abbey Ironworks 27, 47
Newcomen, Thomas 9, 11

Parliament 24-5
Philosophical Magazine 20, 25, 27-8, 29
Plegometer 46
Pole, William 36, 46
Portsmouth Waterworks 57
Price, Joseph 27
Prout, William M. 53

Rainfall 48-50
Registrars *see* Reporters
Reporters (men)
    characteristics 79-80
    job title 22, 41-2
    work 50, 62-4, 76-80, 83
Royal Cornwall Polytechnic Society 37, 50
Rule, John 73

Sims, James 59
Sims, William 28, 50
Smeaton, John 9-11, 81
Stamps 24, 62, 78

Taylor, John
    employs Browne 59
    employs Woolf 22, 29, 30, 37, 40

engine trials 29-30, 40

Historical Statement 22-3, 40-2
in California 75
in Derbyshire 69
in Mexico 7, 55, 73-5
in Spain 75
in Wales 7, 55, 68-71
in Yorkshire 71-2
Lean, Joel (younger) 42-3
Lean's Reporter 37-43
management style 40, 69, 73,
    74-5
Mexican Reporters 73-5
Treffry, J. T. 61, 65
Welsh Reporters 69-71
writings 22, 34, 39-40, 42, 45-6
Telford, Thomas 47
Tilloch, Alexander 19, 20, 25-7 *see
    also* 'Philosophical Magazine'
Tonkin, William 7, 78, 79, 80
Tonkin's Reporter 39, 55, 56-8, 68, 81
Treffry, J. T. 59-61, 65, 67
Trestrail, Nicholas 67-8
Trevithick, Richard 13, 15, 21
    attacks Price 27
    commemorated 53
    on Woolf 33
    plunger poles 13
    praised 42
Trials
    Abraham 24, 28
    Alfred 29
    Consols 29, 30, 40
    Fowey Consols 46, 59
    Towan 29, 71
Vivian, Andrew 21, 25

Water meter 37-9
Watt, James 9-12, 81, 84
West, John 61
West, William
    Browne's Reporter 55, 59, 64-5
    career 65-6
    Fowey Consols 46, 60-1
    Lean 61
    R.C.P.S. 50
Whim engines 15, 24, 78
Whitehurst John 84
Wilson, Thomas 11
Woolf, Arthur
    and Reporters 18-19, 20-2, 25-7,
        31-3, 40
    career 19-21, 22, 25, 37, 42
    character 29, 33
    death 19, 37
    dishonesty 22-3, 30, 31
    engines:
        Abraham 24, 27, 28
        Consols 29, 31
        Fortune 24
        United 29
        Vor 25.27
    Farey, John 20. 28-9, 33
    finances 19-21, 29, 33, 37, 40
    Historical Statement 40, 42
    Leans 21, 22, 31, 37
    Neath Abbey Ironworks 27, 47
    Parliament 25
    patents 19, 29
    pension 40
    premium 20
    returns to Cornwall 18-21
    Taylor 22, 29. 30, 37, 40
    Tilloch 19, 20, 25-7